Andre Schneider

Quantum Sensing Experiments with Superconducting Qubits

Experimental Condensed Matter Physics
Band 29

Eine Übersicht aller bisher in dieser Schriftenreihe erschienenen Bände finden Sie am Ende des Buchs.

Quantum Sensing Experiments with Superconducting Qubits

by
Andre Schneider

KIT Scientific Publishing

Karlsruher Institut für Technologie
Physikalisches Institut

Quantum Sensing Experiments with Superconducting Qubits

Zur Erlangung des akademischen Grades eines Doktors der Naturwissenschaften von der KIT-Fakultät für Physik des Karlsruher Instituts für Technologie (KIT) genehmigte Dissertation

von M. Sc. Andre Schneider

Tag der mündlichen Prüfung: 31. Januar 2020
Referent: Prof. Dr. Martin P. Weides
Korreferent: Prof. Dr. Alexey V. Ustinov

Impressum

KIT Scientific Publishing

Karlsruher Institut für Technologie (KIT)
KIT Scientific Publishing
Straße am Forum 2
D-76131 Karlsruhe

KIT Scientific Publishing is a registered trademark
of Karlsruhe Institute of Technology.
Reprint using the book cover is not allowed.

www.ksp.kit.edu

Print on Demand 2021 – Gedruckt auf FSC-zertifiziertem Papier

ISSN 2191-9925
ISBN 978-3-7315-1032-1
DOI 10.5445/KSP/1000118743

Contents

1 Introduction

"I think there is a world market for maybe five computers."

This famous quote from 1943 is allegedly ascribed to Thomas Watson [1], chairman of IBM, and represents – whether or not original – the expert's view on the business of computers at that time. Filling whole floors with thousands of heat generating vacuum tubes, while still having less computation power than a today's smart watch, only very special applications for computers were imaginable at that time.

But already in 1965, the situation had changed completely, leading Gordon Moore, later CEO of Intel, to the prediction that the structure density on computer chips will double every year [2]. This self-fulfilling prophecy has been driving a tremendous development in computer science and technology for more than 50 years [3, 4]. Decreasing structure sizes and shorter signal paths allow for increasing the computation frequency, but this race stopped in the early 2000s, as the dissipated power density exceeded the one of a hot plate [5]. In addition, the miniaturization itself starts slowing down [6] since computer industry approaches a size where quantum effects, i.e., the quantization of electron charges and tunneling effects, become relevant. When the level of several atoms per building block is reached, the miniaturization comes to a hard end and other ways to improve the computation power have to be found [7].

The demand for more computation power on the other hand is continuing to grow even stronger [8] and the number of known unsolved problems will not stop increasing. Among such problems are the understanding of fundamental biochemical processes like the cycle of photosynthesis or the biological effects of pharmaceutical drugs [9]. Including large molecules and the influence of quantum effects [10], these systems cannot be simulated classically on conventional supercomputers within reasonable time. Richard Feynman therefore proposed a different approach already in 1982:

"nature isn't classical, dammit, and if you want to make a simulation of nature, you'd better make it quantum mechanical" [11]

This idea represents the birth of quantum simulation, where real quantum systems, so far inaccessible for a further investigation, are mapped onto controllable quantum systems, composed of artificial atoms. These man-made quantum systems can be tailored to the specific application, which in turn allows us to gain information about the real-world quantum systems.

Since then, the field of quantum simulation has been evolving and quantum algorithms like Grover's search algorithm [12] and Shor's algorithm for prime factoring [13] gave additional impetus to the field. They promise a faster and more intelligent way of finding the solution by transferring classical problems to the quantum world, where superposition states and entanglement can be exploited. Allowing applications beyond the original idea of quantum simulations, these algorithms created the research field of quantum computation. Here, many conceptual analogies to classical computing are drawn, such as quantum bits (qubits) for the smallest information unit and quantum gates for calculation operations. These analogies allow us to close the circle to the opening quotation of Thomas Watson. Now, instead of vacuum tubes, big cryostats and huge efforts to control single qubits are necessary and so far, consumer applications are again inconceivable.

The computation power of current quantum computing devices is still very limited, which is well represented by a recently demonstrated programmable quantum processor with 53 qubits in a superconducting architecture [14]. It is able to simulate an academic problem in 200 seconds, for which a classical computer would take at least 2.5 days [15]. This achievement is a unique milestone in quantum computing and demonstrates the advance of superconducting quantum systems against other physical implementations.

Already since the first realization of a superconducting qubit in 1999, one of the main challenges in the field has been the fight against decoherence [16]. Although the coherence times have greatly increased by the invention of different qubit designs [17–19] and the improvement of coherence times is already seen in analogy to Moore's law [20], the threshold for useful fault-tolerant quantum computing is not yet reached. This means that despite the experimental demonstration of quantum error correction schemes [21–28], the additional number of qubits and gate operations, which is required to perform this error handling, is still too large to be of practical use.

Fighting decoherence is in particular challenging due to the great intrinsic sensitivity of qubits against a coupling to other systems. Since qubits are true quantum systems, the qubit state will be projected onto one of two states, as soon as somebody "looks", i.e., when the qubit couples to a different, observable system. Especially for macroscopic quantum systems, the freedom of designs and the possibility to choose their properties in a great parameter space come at the cost of increased parameter fluctuations, which results in additional noise and decoherence. One major goal is therefore to find the most prominent spurious coupling mechanisms and systems, and follow new paths to circumvent their influence.

On the way to this goal, new systems and effects are found to be relevant and are studied in greater detail. Therefore, the vulnerability of a quantum state and the extreme sensitivity of a qubit can also be seen as unique feature, enabling the investigation of other systems by means of a very sensitive sensor. Together with other mechanisms exploiting quantum effects, the basis for the field of quantum sensing is formed.

Although many sensing experiments have been performed on other physical systems, superconducting qubits are again promising candidates, as they allow for an easy coupling due to their large magnetic and electric dipole moments [29]. Through techniques like dynamical decoupling [30] or the evaluation of consecutive Ramsey experiments [31], they can be used to detect noise in a tunable frequency range from the millihertz to the gigahertz regime. It is also possible to discriminate between magnetically and electrically induced noise on the same device by changing the bias point of phase and flux qubits [32, 33]. Via relaxometry studies, superconducting qubits have been used to demonstrate the presence of single atomic fluctuators, whose frequency can be tuned by electric fields and mechanical strain, and which heavily influence the qubit coherence properties [34–37]. These results of quantum sensing experiments therefore assist the development of more coherent qubits, creating a strong link between the fields of quantum sensing with superconducting qubits and quantum computation.

This link is even more obvious in the development of a quantum computer, where many qubits have to be combined onto a single quantum chip, together with a vast amount of control and auxiliary structures. This variety of circuit details leads to an increased number of coupling partners and therefore loss channels to the qubit. Although microwave circuits can be classically simulated by numeric methods, these simulations do not include the nonlinearity of the qubits, imperfections in the fabrication process and experimental inaccessibilities. A precise experimental calibration of the cross-talk between individual components as well as a characterization of the microwave transmission from the control electronics to each qubit individually is therefore of great interest.

To enable this calibration, a sensing scheme is derived in this thesis by utilizing the AC Stark shift of a multi-level superconducting circuit. This is then demonstrated by measuring the frequency-dependent transmission of microwave signals from room temperature devices to the quantum bit. This sensing scheme is additionally of great interest for the development of new hybrid quantum systems, where the harmonic resonance of the second system could be detected with the qubit. Such hybrid quantum systems have gained much interest in research as they are able to combine the rapid processing of superconducting qubits with, e.g., the long coherence times of paramagnetic spin ensembles [38–42] or nanomechanical oscillators for ultra-sensitive vibrational sensing [43, 44]. In combination with micromechanical and magnetooptical devices, an interface to optical photons can be established [45, 46], which is in particular relevant for quantum information processing. Here, it would enable to coherently transfer quantum states between remote quantum computing units and allow for a secure quantum communication over long distances [47, 48]. In general, hybrid quantum systems can be used to combine the benefits of two independent systems [49].

The border between the fields of hybrid quantum systems and quantum sensing is weak when a quantum bit is used to study a second quantum system, and the naming mainly depends on the research direction, even if the physical realization is identical. In quantum sensing, typically a known quantum system (i.e., the sensor) is used to study another, unknown (quantum) system. For a hybrid quantum system instead, the focus often lies on application, where benefits are combined and unique features appear [49]. Two prominent examples for a hybrid quantum system are used throughout this thesis: First, the qubit forms a hybrid system with a superconducting resonator. In this scheme, the resonator frequency exhibits a dispersive shift depending on the qubit state, meaning that the qubit state can be inferred by probing the resonator. At the same time, the resonator acts as band-pass filter for the qubit, protecting it from unwanted microwave radiation and excitation loss. Second, we study a qubit-magnon hybrid system, which can be used in future studies to create a link between stationary superconducting qubits and flying optical qubits, to enable quantum communication.

Realizing the coupling between a qubit and the collective excitations in a ferromagnetic material would drive the highly active field of magnonics [50, 51] into the quantum regime, allowing for a more fundamental understanding of the relevant mechanisms. In contrast to electronics and spintronics, where the charge and spin of moving electrons are used as information carrier, magnonics relies on the magnetic exchange interaction between neighboring spins and hence realizes the transport of information without a physical movement of charged particles. Therefore, magnonic devices circumvent several problems of modern electronics, like heat dissipation by Ohmic losses, as demonstrated by the implementation in ferrimagnetic insulators [52]. With interference-based components [53], the footprint of computation elements can be significantly decreased compared to standard CMOS[1] technology, new multi-port gates can combine several logical elements into one [54], and frequency multiplexing up to the terahertz regime [55] allows for a massive intrinsic parallelization of computation tasks [56]. With effects like the distortion-free propagation of magnon solitons [57] and the Bose-Einstein condensation at room-temperature [58, 59], interesting physical phenomena exist, demanding for further research and investigation.

Although magnon spin waves have been predicted in 1923 [60] and have been subject to research and applications since the middle of the last century [61], most studies have been performed with high powers at room temperature. Magnon research at

[1] complementary metal oxide semiconductor, the standard in electronics

low temperatures and low excitation power has only recently emerged, opening new possibilities to approach and study the spin system's ground state [62–65]. A true step into the quantum regime can however only be reached when implementing a nonlinearity in the system. This would then allow for magnon counting and a coherent manipulation of the magnonic state [66–69].

Again, superconducting qubits suggest themselves for quantum sensing experiments with magnons in the low-excitation regime, as they are designable macroscopic quantum objects which operate at millikelvin temperatures in the gigahertz regime. In this thesis, we therefore place a ferromagnetic material in close vicinity to a superconducting qubit, to study their interaction and the properties of magnon excitations. However, this implies several experimental challenges, as the magnetic material needs an external bias field, while superconductivity is very susceptible to magnetic fields. State-of-the-art implementations therefore separate qubit and magnetic system spatially by adding a coupling cavity [66]. This only creates a cavity-mediated coupling instead of a direct interaction, which adds additional couplings and complexity. Therefore, we start to explore the experimentally challenging way of a direct qubit-magnon coupling in this thesis. Using a planar approach, the magnonic system is placed directly on top of the qubit, meaning that the qubit is operated in a magnetic field.

Since no experimental data on the compatibility of superconducting qubits with magnetic fields are available, we investigate this dependency in the framework of this thesis, and find that a quantum coherent behavior persists up to magnetic fields far beyond what has been expected previously [70]. Besides these promising results, the acquired data reveal more insights about the qubit's internal material structure and the phase coherence of the qubit state.

Based on these measurements, different magnetic materials are considered for the hybrid circuit and new theoretical descriptions and experimental methods to realize the coupling are investigated. Since the magnetic material of choice was not studied at low temperatures before, the mode spectrum and internal losses were unknown. To this end, the previously introduced sensing scheme can be applied, which enables the qubit to detect harmonic signals.

In the following chapters, I first provide the theoretical background to understand the basic function of a superconducting qubit and the expected behavior of a superconductor in magnetic fields. After a summary on the measurement techniques in the microwave regime, the measurement setup and samples are presented. The subsequent discussion of experimentally acquired data is divided into three parts: First, the new quantum sensing scheme based on the AC Stark effect is described, which exploits the presence of higher excitation levels in the qubit. With that, the sensitivity is increased and a simultaneous measurement of amplitude and frequency is enabled [71, 72]. The application of this sensing scheme is demonstrated by characterizing the frequency-dependent microwave transmission from the measurement lines to the location of the qubit. Afterwards, the performance of a qubit-resonator system in a magnetic field is characterized and discussed. These data show promising results for further applications and reveal the relevance of details in the qubit's fabrication process [70]. After a summary of the theoretical background for collective magnetic excitations, the coupling scheme between qubit and magnetic material is elaborated and the behavior of the magnon-qubit hybrid system is characterized. Here, a rich mode spectrum is found, giving a promising outlook on future experiments and applications.

2 Superconducting quantum bits

In this chapter, I provide a summary of the theoretical background necessary to understand the experiments with superconducting quantum bits presented in this work. Since many topics and effects have to be discussed and mentioned here to enable the reader to understand the experimental data, I tried to simplify the following explanations as much as possible. Therefore, some statements might look a bit over-simplified for the experienced reader, but they retain the key features of our quantum systems.

This chapter starts with a general theoretical discussion of quantum bits, which is valid for all kinds of physical realizations and broadens up to multi-level quantum systems. After an introduction to superconductivity, these two topics are merged into a section about the superconducting transmon qubit, its features and limitations.

2.1 Theory of quantum bits

2.1.1 Qubit states and the Bloch sphere

In analogy to a classical bit, a quantum bit, or qubit for short, has two possible eigenstates, $|0\rangle$ and $|1\rangle$. Due to its quantum nature, it can also be in a superposition of these states, resulting in an arbitrary qubit state of

$$|\Psi\rangle = a\,|0\rangle + b\,|1\rangle$$

$$|0\rangle = \begin{pmatrix} 0 \\ 1 \end{pmatrix}, \quad |1\rangle = \begin{pmatrix} 1 \\ 0 \end{pmatrix}, \quad |\Psi\rangle = \begin{pmatrix} b \\ a \end{pmatrix}.$$

The complex amplitudes a and b have to fulfill $|a|^2 + |b|^2 = 1$, as $|\Psi\rangle$ has to be normalized with $\langle\Psi|\Psi\rangle = 1$. The global phase of $|\Psi\rangle$ has however no physical meaning, so we can chose $0 \leq a \leq 1$, $a \in \mathbb{R}$ without loss of generality. These restrictions are intuitively implied by spherical coordinates, with azimuthal angle θ and polar angle ϕ:

$$|\Psi\rangle = \cos\frac{\theta}{2}\,|0\rangle + \mathrm{e}^{\mathrm{i}\phi}\sin\frac{\theta}{2}\,|1\rangle\,. \tag{2.1}$$

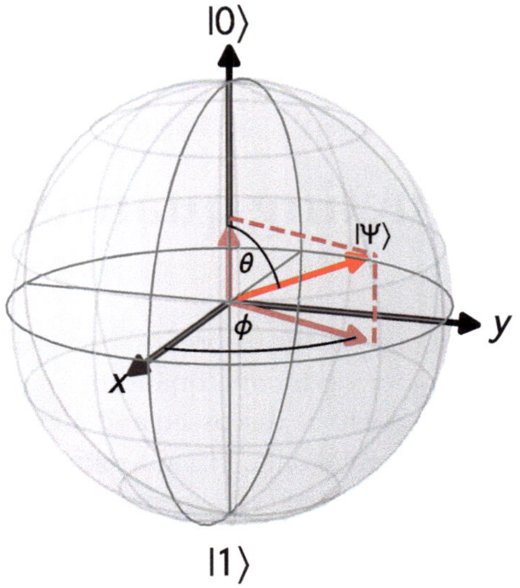

Figure 2.1: Bloch sphere representation. An arbitrary qubit state $|\Psi\rangle$ on the Bloch sphere can be described by the two angles θ and ϕ known from spherical coordinates. For a pure quantum state, the state is always on the surface of the Bloch sphere, $||\Psi\rangle| = 1$. The axes are chosen such that $(|0\rangle + |1\rangle)\sqrt{2}$ is on the x axis and $(|0\rangle + \mathrm{i}\,|1\rangle)\sqrt{2}$ points along the y axis.

As these points lie on the surface of a sphere, this gives the basis for representing an arbitrary qubit state in the Bloch sphere, as illustrated in Fig. 2.1. The common choice of aligning $|0\rangle$ along the positive z-axis may seem counterintuitive at first, but has historical reasons related to experiments with a spin in a magnetic field. There, the bias field is intuitively chosen to be aligned in positive z direction resulting in the spin in the ground state also points upwards. Although a great number of qubit architectures do not require a biasing field nowadays, this convention persists.

2.1.2 Qubit operators

Any operation on a qubit state can be expressed by a linear combination of the Pauli rotation matrices $\hat{\sigma}_x$, $\hat{\sigma}_y$ and $\hat{\sigma}_z$. In the Bloch sphere representation, each matrix operation corresponds to a rotation around the respective axis by an angle of π. One common combination of rotation operators is the creation/annihilation operator

$$\sigma_{\pm} = \frac{1}{2}\left(\hat{\sigma}_x \pm \mathrm{i}\hat{\sigma}_y\right), \tag{2.2}$$

which excites the ground state $\sigma_+ |0\rangle = |1\rangle$ and deexcites the excited state $\sigma_- |1\rangle = |0\rangle$. The excitation probability of the qubit can be acquired by measuring the expectation value of the $\hat{\sigma}_z$ operator:

$$\langle \hat{\sigma}_z \rangle = \langle \Psi | \hat{\sigma}_z | \Psi \rangle = -|a|^2 + |b|^2 = -\cos\theta = -1 \ldots 1. \tag{2.3}$$

For a qubit transition energy of $E = \hbar\omega_{01}$, this results in a qubit Hamiltonian of

$$H_\mathrm{q} = \hbar\frac{\omega_{01}}{2}\hat{\sigma}_z. \tag{2.4}$$

With the time evolution operator given by $U = \exp -\frac{\mathrm{i}}{\hbar}H(t - t_0)$, the evolution of the qubit state is given as

$$|\Psi(t)\rangle = U(t)\,|\Psi(t_0 = 0)\rangle = \cos\frac{\theta}{2}\,|0\rangle + \mathrm{e}^{\mathrm{i}(\phi - \omega_{01}t)}\sin\frac{\theta}{2}\,|1\rangle\,.$$

This means that the qubit state always rotates around the z-axis with a speed of ω_{01}, corresponding to a Lamor precession. For a more intuitive understanding, this rotation is typically removed by describing the qubit state in a frame rotating at the frequency of the qubit. Since the tool we use to investigate the qubit in our experiments, i.e., a microwave drive, rotates itself with a frequency ω_D, we naturally shift our perspective to a rotating frame. Consequently, we only see a remaining rotation at the speed of $\omega_{01} - \omega_\mathrm{D}$.

2.1.3 Driving a qubit

To drive the qubit transition, we apply a microwave tone of frequency ω_D, amplitude A and phase offset φ, which results in the drive Hamiltonian

$$\hat{H}_\mathrm{D}/\hbar = A\cos\left(\omega_\mathrm{D}t + \varphi\right)\hat{\sigma}_x.$$

With the unitary transformation

$$\hat{U} = \exp\left(\frac{\mathrm{i}}{2}\omega_\mathrm{D}t\hat{\sigma}_z\right), \tag{2.5}$$

we transform the drive and the qubit Hamiltonian Eq. (2.4) to a frame rotating with the drive frequency

$$\begin{aligned}
\tilde{H}\hbar &= \hat{U}\left(\hat{H}_\mathrm{q} + \hat{H}_\mathrm{D}\right)\hat{U}^\dagger/\hbar - \mathrm{i}\hat{U}\dot{\hat{U}}^\dagger \\
&= \frac{1}{2}\Delta_\omega\hat{\sigma}_z + \frac{A}{2}\left(e^{\mathrm{i}(2\omega_\mathrm{D}t+\varphi)}\hat{\sigma}_+ + e^{\mathrm{i}\varphi}\hat{\sigma}_- + e^{-\mathrm{i}\varphi}\hat{\sigma}_+ + e^{-\mathrm{i}(2\omega_\mathrm{D}t+\varphi)}\hat{\sigma}_-\right) && (2.6)\\
&\approx \frac{1}{2}\Delta_\omega\hat{\sigma}_z + \frac{A}{2}\left(e^{\mathrm{i}\varphi}\hat{\sigma}_- + e^{-\mathrm{i}\varphi}\hat{\sigma}_+\right) && (2.7)\\
&= \frac{1}{2}\Delta_\omega\hat{\sigma}_z + A\left(\cos(\varphi)\hat{\sigma}_x + \sin(\varphi)\hat{\sigma}_y\right) \\
&= \frac{1}{2}\begin{pmatrix} A\cos\varphi \\ A\sin\varphi \\ \Delta_\omega \end{pmatrix} \cdot \begin{pmatrix} \hat{\sigma}_x \\ \hat{\sigma}_y \\ \hat{\sigma}_z \end{pmatrix}, && (2.8)
\end{aligned}$$

where the detuning $\Delta_\omega = \omega_q - \omega_\mathrm{D}$ between drive and qubit frequency was introduced. In a rotating wave approximation, fast rotating terms have been neglected from Eq. (2.6) to Eq. (2.7).

From Eq. (2.8) we see that the phase φ of the microwave signal sets the axis around which we rotate in the xy plane. As this phase can only be defined in relation to a second pulse, and the absolute phase of a quantum state $|\Psi\rangle$ is not of interest, the first microwave pulse on the qubit defines the phase reference. We also see that the last line containing $\hat{\sigma}_z$ does not depend on the amplitude of the drive tone. This is consistent with the qualitative argument given in the previous section, that we only see the frequency mismatch Δ_ω between drive tone and qubit frequency rather than the qubit frequency itself.

For many experiments it is sufficient to consider only a rotation around the $\hat{\sigma}_x$ axis, resulting in $\varphi = 0$ and Eq. (2.8) simplifies to

$$\tilde{H}/\hbar = \frac{A}{2}\hat{\sigma}_x + \frac{\Delta_\omega}{2}\hat{\sigma}_z, \tag{2.9}$$

with the eigenenergies

$$E_{\pm}/\hbar = \pm\frac{1}{2}\sqrt{A^2 + \Delta_\omega^2} = \frac{1}{2}\Omega_\mathrm{R}.$$

Here, Ω_R is the generalized Rabi frequency. From a more exact analysis, we get that for a non-vanishing detuning Δ_ω, the effective rotation axis is no longer the $\hat{\sigma}_x$ axis but a tilted axis. This leads to an increased rotation frequency $\Omega_\mathrm{R} \geq A$ and a decreased amplitude, meaning that the state $|1\rangle$ is never reached by a detuned drive [73].

2.1.4 Dispersive readout of a qubit

For different qubit architectures, different readout techniques can be used, where e.g. the flux close to a flux qubit is measured [74] or the tunneling current in a charge qubit [16]. However, there is a more general approach, similar to the Jaynes-Cummings model [75] used in quantum optics. The qubit (an atom) is coupled to a resonator (a cavity) and the system is probed by applying photons to the resonator. The Hamiltonian for this system is given by

$$\begin{aligned}\hat{H} = &\quad \hat{H}_\mathrm{res} \quad + \quad \hat{H}_\mathrm{q} \quad + \quad \hat{H}_\mathrm{int} \\ = &\ \hbar\omega_\mathrm{r}\left(\hat{a}^\dagger\hat{a} + \frac{1}{2}\right) + \frac{\hbar}{2}\omega_{01}\hat{\sigma}_z + \hbar g\left(\hat{\sigma}_+\hat{a} + \hat{\sigma}_-\hat{a}^\dagger\right).\end{aligned} \tag{2.10}$$

Here, ω_r denotes the resonator frequency with its photon creation and annihilation operators $\hat{a}^\dagger$ and $\hat{a}$ and g is the coupling strength between qubit and resonator.

Through the mutual coupling, the eigenfrequencies of the system change in dependence of the number of photons in the whole system. In the dispersive limit, where $\Delta_\mathrm{r} = \omega_\mathrm{r} - \omega_{01} \gg g$, we can simplify the expression to [76]

$$\hat{H}_\mathrm{eff}/\hbar \approx (\omega_\mathrm{r} + \chi\hat{\sigma}_z)\,\hat{a}^\dagger\hat{a} + \frac{1}{2}(\omega_{01} + \chi)\,\hat{\sigma}_z \tag{2.11}$$

where terms of the order $\mathcal{O}(g^2/{\Delta_\mathrm{r}}^2)$ are neglected in a perturbative approach. With the dispersive shift χ, given for true two-level systems as $\chi = g^2/\Delta_\mathrm{r}$, we get an effective resonance frequency of the resonator of $\omega_r + \chi\hat{\sigma}_z$. Measuring the resonance frequency, we perform a projective measurement of $\hat{\sigma}_z$, i.e., we measure the qubit state. Since the wavefunction $|\Psi\rangle$ of the qubit collapses into one of the eigenstates of the $\hat{\sigma}_z$ operator, we get for each measurement one of the two possibilities $\omega = \omega_r \pm \chi$. To measure the probability $P_1(\Psi) = |\langle 1|\Psi\rangle|^2$, we need to repeat this projective measurement and apply statistics.

The expectation values $\langle\hat{\sigma}_x\rangle$ and $\langle\hat{\sigma}_y\rangle$, i.e., the projection of the Bloch vector onto the xy plane cannot directly be accessed for our qubit types. However, by applying a $\hat{\sigma}_y/2$ or $\hat{\sigma}_x/2$ rotation pulse directly before the dispersive readout, these values also become accessible [77].

2.1.5 Mixed states

So far, we only considered pure states with $||\Psi\rangle|^2 = 1$ which are located on the surface of the Bloch sphere. Under the influence of spontaneous relaxation via the $\hat{\sigma}_-$ operator or dephasing via random fluctuations of ω_{01}, our statistically measured state is no longer a pure state. This mixed state can no longer be described by a single ket, but by means of a density matrix:

$$\hat{\rho} = \sum_i p_i |\Psi_i\rangle \langle\Psi_i| , \qquad (2.12)$$

where p_i is the fraction of the ensemble state being in state $|\Psi_i\rangle$. These mixed states are represented by points within the Bloch sphere, which have a shorter distance to the center of the sphere.

More detailed descriptions of the relaxation and dephasing processes are given in Sec. 4.2.3, where the measurement procedure is explained.

2.2 From qubits to qutrits and qudits

In the theoretical model of a quantum bit, only two qubit eigenstates, namely $|0\rangle$ and $|1\rangle$ are considered, disregarding the existence of further energy levels. In the vast majority of physical realizations for quantum bits however, more energy levels exist and are treated very differently: If these energy levels are far away in terms of transition energy, they usually are ignored, e.g. in quantum bits using atomic transitions or a spin degree of freedom. In other systems, like nitrogen vacancy (NV) centers in diamond, auxiliary energy levels are used to drive conditional transitions which favor the population of one of the qubit levels or they are used to read out the state of the qubit by photoluminescence [78].

Additional energy levels can also be used for special computation or simulation algorithms, if the transitions between these levels can be addressed individually. Grover's algorithm [12] for the search in a large unstructured (quantum) database for instance was theoretically modified [79] to operate on multiple energy levels of a single physical system instead of many independent qubits. This method was recently demonstrated on the four energy levels of a single magnetic molecule's nuclear spin [80].

Since the name qu**bit** is reserved for two-level quantum systems, the names qu**trit** for a three level quantum system and qu**dit** for general multi level quantum systems have been introduced. However, not every multi-level quantum system can be seen as a valid qudit as all levels need to be individually addressable and controllable.

This is possible if the system features an anharmonicity, which makes the transition frequencies between the levels distinguishable.

Within certain limitations it is possible to ignore the higher energy levels of a qudit and treat it as a true qubit, which simplifies the mathematical expressions and theoretical description. To follow common terminology, throughout this work multi-level systems will be called "qubits" as long as only the first two energy levels are actively used, while higher levels are only considered to explain spurious effects and limitations. When more than two levels are required to describe an effect, the term "qudit" is employed, even if the same system was previously referred to as qubit.

Many different physical systems are used to realize qubits or qudits, each having its own benefits and drawbacks. For this work, superconducting qubits have been chosen as they are macroscopic quantum objects that are well designable and simulatable, and offer different coupling mechanisms. Since there are schemes to integrate several qubits on one chip and their fabrication methods are closely related to the well-established procedures used in the semiconducting industry, they are one of the most promising systems to build a scalable quantum computer. So far they are the first and only platform, where quantum supremacy, the benefit of quantum computing against classical computation, was demonstrated [14].

In the next sections, the fundamentals of superconducting qubits are introduced.

2.3 Superconductivity

Superconductivity was first discovered by the Dutch physicist Heike Kammerlingh Onnes in 1911 [81] and described as a sudden drop in resistivity of pure mercury at very low temperatures. Superconductors are therefore ideal conductors and can carry a current without any dissipation. Several years later, it was discovered that superconductors are not only ideal conductors but also expel any magnetic field from their inside as soon as they are cooled below their critical temperature, the so-called Meißner-Ochsenfeld effect [82]. Hence, superconductors also are ideal diamagnets. These unique properties are however subject to limited validity, resulting in a critical temperature T_c, a critical current density j_c and a critical field H_c, above which superconductivity breaks down.

Apart from many phenomenological descriptions, no fundamental theory for the underlying reasons to superconductivity was found until Bardeen, Cooper and Schrieffer developed the BCS theory [83]. The key to this theory is a phonon-mediated weak attractive force between two electrons, which leads – despite their Coulomb repulsion – to the formation of electron pairs, known as Cooper pairs. Pairing two spin 1/2 particles results in a particle of integer spin, which is therefore of bosonic nature. Bosons in turn do not follow Pauli's exclusion principle [84] and the Cooper pairs condensate into the BCS ground state, which is described by a single macroscopic wave function

$$\Psi(\vec{r}) = |\Psi| \, \mathrm{e}^{\mathrm{i}\theta(\vec{r})}, \tag{2.13}$$

with the macroscopic phase $\theta(\vec{r})$. This wave function spreads over the whole condensate, i.e., the whole superconductor. As it does not decay spatially, for any path on a closed superconducting loop we can only accumulate a phase of integer multiples of 2π to maintain a consistent wavefunction. In other words the wavelength of the Cooper pairs has to be an integer fraction of the ring's circumference.

The wave vector $|\vec{k}| = 2\pi/\lambda$ for a charged quantum particle does not only depend on the kinetic energy or momentum of the particle, but also on its surrounding vector potential $\vec{A}$ with $\vec{\nabla} \times \vec{A} = \vec{B}$. We get

$$\hbar\vec{k} = m\vec{v} + q\vec{A}.$$

We now demand that the integration of the wave vector around a closed ring results in an integer multiple n of 2π

$$2\pi n = \oint \vec{k} \cdot \mathrm{d}\vec{r} = \frac{m}{\hbar} \oint \vec{v} \cdot \mathrm{d}\vec{r} + \frac{q}{\hbar} \oint \vec{A} \cdot \mathrm{d}\vec{r} \tag{2.14}$$

and can replace $\oint \vec{A} \cdot \mathrm{d}\vec{r} = \int \vec{\nabla} \times \vec{A} \cdot \mathrm{d}\vec{S} = \int \vec{B} \cdot \mathrm{d}\vec{S} = \Phi$ using Stokes' theorem, where Φ is the magnetic flux. This gives

$$\begin{aligned} 2\pi n \frac{\hbar}{q} &= \frac{m}{q} \oint \vec{v} \cdot \mathrm{d}\vec{r} + \Phi \\ n\frac{h}{q} &= \frac{m}{q^2 n_\mathrm{s}} \oint \vec{j_\mathrm{s}} \cdot \mathrm{d}\vec{r} + \Phi, \end{aligned} \tag{2.15}$$

where the supercurrent density $\vec{j_\mathrm{s}} = q n_\mathrm{s} \vec{v}$ was introduced together with the density of Cooper pairs n_s. For a superconductor with thick walls, we can assume that the shielding currents are only flowing in a thin layer below the surface and that a path with $\vec{j_\mathrm{s}}(\vec{r}) = 0$ exists: in this case the integral vanishes and we finally find the flux quantum Φ_0 to be

$$\Phi_0 = \frac{\Phi}{n} = \frac{h}{q} = \frac{h}{2e}, \tag{2.16}$$

where $q = 2e$ for the Cooper pairs was used. This means that the flux flowing through a hole in a superconductor is quantized and can only be present in integer multiples of Φ_0.

If the superconductor however is free of holes, only $n = 0$ can be the solution to Eq. (2.14), as $\vec{k}(\vec{r})$ is continuous. From Eq. (2.15) we hence get

$$\frac{m}{q^2 n_s} \oint \vec{j}_s \cdot d\vec{r} = -\Phi = -\int \vec{B} \cdot d\vec{S}.$$

Using again Stokes' theorem and comparing the integrand, this results in

$$\vec{B} = -\frac{m}{q^2 n_s} \vec{\nabla} \times \vec{j}_s. \tag{2.17}$$

Taking the curl of both sides in Ampère's circuital law $\vec{\nabla} \times \vec{B} = \mu_0 \vec{j}_s$, where a non-magnetic superconductor ($\mu = 0$) and $\vec{j} = \vec{j}_s$ is assumed, we get

$$\mu_0 \vec{\nabla} \times \vec{j}_s = \vec{\nabla} \times \left(\vec{\nabla} \times \vec{B}\right) = \vec{\nabla}\left(\vec{\nabla} \cdot \vec{B}\right) - \vec{\nabla}^2 \vec{B} = -\vec{\nabla}^2 \vec{B}, \tag{2.18}$$

where Maxwell's equation $\vec{\nabla} \cdot \vec{B} = 0$ was used. Combining now Eq. (2.17) and Eq. (2.18) gives

$$\vec{\nabla}^2 \vec{B} = \frac{\mu_0 q^2 n_s}{m} \vec{B}$$

or, regarding a superconducting semispace in $x > 0$ and vacuum outside

$$\frac{d^2 B_z(x)}{dx^2} = \frac{\mu_0 q^2 n_s}{m} B_z(x)$$

resulting in the solution

$$B_z(x) = B_z(0) e^{-x/\lambda_L} \qquad \text{with } \lambda_L = \sqrt{\frac{m}{\mu_0 q^2 n_s}}. \tag{2.19}$$

This means that the magnetic field does not discontinuously drop to zero but decays exponentially from the surface of the superconductor on the length scale of the London penetration depth λ_L as we go deeper into the superconductor. This is depicted in Fig. 2.2 a). With $\vec{\nabla} \times \vec{B} = \mu_0 \vec{j}_s$ the screening currents decay on the same length scale.

2.3.1 Type I and type II superconductors

Expelling the magnetic field from the inside of the superconductor costs energy and is therefore only possible up to certain fields: Once the energy needed to start the screening current and provide the compensation field is higher than the energy gained by the condensation of the Cooper pairs, the superconducting state is no longer

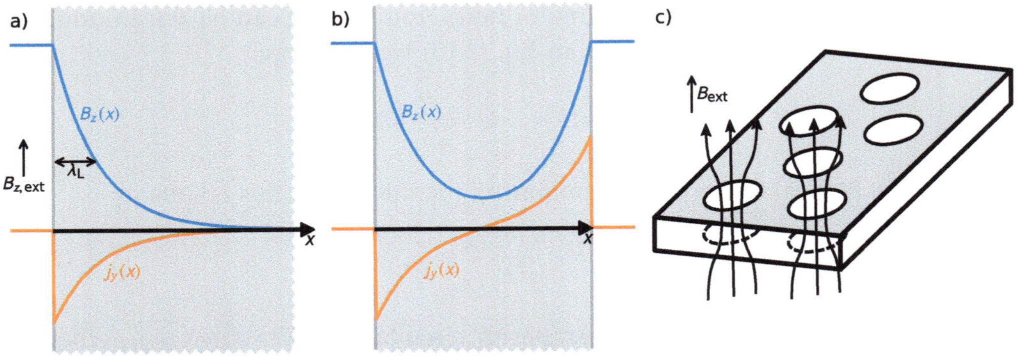

Figure 2.2: Boundary effects of superconductors. Magnetic field distribution at the surface of a superconductor. a) An infinte superconducting semispace for $x > 0$ in an outer magnetic field. The field decays over a length of λ_L and screening currents flow at the surface. b) A thin superconductor in a parallel field. The magnetic field inside the superconductor never approaches zero, which requires less energy for the expulsion of the field. This configuration can therefore be superconducting up to higher fields. c) A type II superconductor in a perpendicular magnetic field. Normal conducting cores are created, through which flux can flow. They repel each other and align in a hexagonal lattice to have the largest distance.

favorable and superconductivity breaks down. Therefore, a critical magnetic field[1] B_c exists, beyond which the superconductor turns normal.

For different geometries and materials it is possible that not the whole superconductor turns normal at the same time, but only parts get normal conductive. In this regime, a third energy comes into play: the energy of the interface between superconductor and normal conductor. To describe this interface, we use the formalism of the Ginzburg-Landau equations [85] and focus on a qualitative explanation of the results. Detailed derivations can be found in any standard physics textbooks about superconductivity [86–88]. To describe the interface, the London penetration depth λ_L from Eq. (2.19) is relevant together with the Ginzburg-Landau coherence length ξ_GL, which can be intuitively understood as the spatial extent of a Cooper pair. Due to the strong correlation of the Cooper pairs it is clear that any variation in the density of Cooper pairs n_s can only occur on the length scale of ξ_GL. In the normal conducting area it has to be $n_\mathrm{s} = 0$. In simple words, the superconductivity builds up over the length scale ξ_GL from the boundary, meaning that the total gain in condensation energy is decreased with increasing ξ_GL and increasing area of the interface. On

[1] The terms "magnetic field" and "magnetic flux density" are used synonymously throughout this thesis, following common usage. Whenever a clear differentiation is needed, it will be stated by the used quantity symbols, H and B.

the other hand, the energy needed to expel the flux the superconductor decreases, when λ_L is high and the interface is large.

Eventually, this means that for $\lambda_L/\xi_{GL} \ll 1$ the creation of boundaries costs energy and is not favorable, and such a superconductor turns normal at once. This is a **type I** superconductor. For the case $\lambda_L/\xi_{GL} \gg 1$, boundaries become energetically favorable at a certain magnetic field $B_{c,1}$ and magnetic flux starts to enter the **type II** superconductor. To maximize the number of boundaries, the flux inside the normal conducting area has to be as small as possible, which means after Eq. (2.16) that flux vortices containing one flux quantum Φ_0 are created, until the second critical field $B_{c,2}$ is reached, where the superconductivity eventually breaks down. The state between $B_{c,1}$ and $B_{c,2}$ is called the vortex state and depicted in Fig. 2.2 c).

The ratio λ_L/ξ_{GL} can be influenced by the mean free path length l of conduction band electrons in the normal conducting material. While λ_L slightly increases with decreasing l, ξ_{GL} decreases strongly. It is therefore easily possible to decrease the mean free path length by introducing defects or doping with other metals and with that turn a type I into a type II superconductor. This relation is usually expressed in the form that a superconductor in the "dirty limit" is always a type II superconductor.

2.3.2 Thin films in a magnetic field

So far, only the influences of magnetic fields on large superconducting samples have been discussed. If we now apply a magnetic field in **parallel** to a thin superconducting film ($d \approx \lambda_L$), the field enters from both sides into the superconductor, the screening currents cancel out in the center and the superconductor is not at any point free from magnetic fields, as shown in Fig. 2.2 b). This behavior can again be described and modeled by the Ginzburg-Landau equations [85] and in-depth descriptions can be found in original research publications [89] as well as in standard physics textbooks [86–88]. Due to a large number of degrees of freedom in the experiments presented in Ch. 6 and Ch. 7, where a superconductor in an in-plane magnetic field is studied, the derivation of an exact formula here would bring little additional insight. It shall hence be sufficient for us to stick to the following qualitative argumentation:

With the magnetic field entering from both edges of the superconductor, a considerable magnetic field is still present at the center of the sample. As already discussed, expelling the magnetic field from the inside of the sample is (disregarding other interactions) energetically not favorable. This intuitively becomes clear when comparing the superconductor to a classical diamagnet where all molecular magnets have to align opposite to the outer magnetic field. This means that the magnetic moment per unit volume is reduced for a thin film compared to the bulk case and a thin film (i.e. a smaller magnetic needle) is more stable in an outer magnetic field and can withstand higher fields, before it "flips". Eventually, this leads to the experimentally proved conclusion that the critical parallel magnetic field for a thin superconducting film is much higher than for the bulk material.

For the case of a thin superconducting film in a **perpendicular** magnetic field, the relevant length scale is not λ_L but again the Ginzburg-Landau coherence length ξ_{GL}. While thicker films with $d \gg \xi_{GL}$ form characteristic patterns of normal conducting areas, the so-called intermediate state, thinner films of type I superconductors, where $d \ll \xi_{GL}$, behave just like regular type II superconductors [90] with the creation of flux vortices containing a single flux quantum. The qualitative argumentation here is similar to the limit of a dirty superconductor where now the thickness of the sample restricts the mean free path and coherence length. In the intermediate regime $d \approx \xi_{GL}$ however, this approximation is very crude and more sophisticated thoughts have to be taken [90]. For our experiments, we can still restrict our thoughts to the limit of thin superconductors[2].

2.4 Josephson Effect

If two superconductors are connected by a weak link, e.g., a thin normal metal or insulator, the Cooper pairs can tunnel coherently through this barrier and the wavefunctions of the two superconductors overlap. Despite the insulating barrier, the current can still flow without resistance, which is called the Jopsephson effect [92]. Therefore, no voltage drop occurs, but the superconducting phase changes, i.e., there is a phase drop across the junction. The gauge independent phase difference between a position before (1) and after (2) the junction,

$$\varphi = \theta_1 - \theta_2 - \frac{2\pi}{\Phi_0} \int_1^2 \vec{A} d\vec{l}, \tag{2.20}$$

is related to the current flowing through the junction via

$$I(\varphi) = I_c \sin \varphi, \tag{2.21}$$

where I_c is the critical tunnel current of the junction. From the derivative of the gauge invariant phase, we get the equation

$$V = \frac{\hbar}{2e}\dot{\varphi} = \frac{\Phi_0}{2\pi}\dot{\varphi}, \tag{2.22}$$

[2] For aluminum, $\xi = 1.6\,\mu\text{m}$ was measured [91].

meaning that the voltage across the junction is proportional to the change in phase and therefore dependent on the change in current. Compared to the voltage at a classical inductance $V = L\dot{I}$, we can define the Josephson inductance

$$L_{\mathrm{J}}(\varphi) = \frac{\Phi_0}{2\pi I_{\mathrm{c}}} \frac{1}{\cos\varphi}. \tag{2.23}$$

A Josephson junction (JJ) can therefore be seen as a nonlinear inductance, where the inductance depends on the current flowing.

This nonlinear element enables us to build an anharmonic quantum oscillator and use it as a qubit, which will be introduced in Sec. 2.5. For the moment, we will have a short look on a different application and properties of Josephson junctions.

2.4.1 The SQUID

If we interrupt a superconducting loop by two JJs, as depicted in Fig. 2.3 a) it is clear that after Eq. (2.15), the flux Φ enclosed in the loop can only take integer multiples of Φ_0. This means that screening currents in the loop start to circulate when an external magnetic flux Φ_{a} is applied. If we now drive an external current I_{b} across the junctions, their effective critical current is reduced. This setup is commonly called a SQUID – a superconducting quantum interference device. It can either be used as very sensitive sensors for magnetic fields by increasing the area of the flux loop, or as one effective JJ whose critical current $I_{\mathrm{c,eff}}$ can be tuned by an external magnetic field.

2.4.2 A Josephson junction in a magnetic field

When placing a single JJ in an in-plane magnetic field, depicted in Fig. 2.3 b), the field enters preferably in the regions of lower critical current, i.e., the junction. This also becomes clear from λ_{L} in Eq. (2.19) when taking into account that the density of Cooper pairs n_{s} is reduced in the JJ. The magnetic field in the junction leads to a gradient in the vector potential $\vec{A}$, giving a contribution to the gauge invariant phase (Eq. (2.20)). We can split the junction into an array of parallel junctions, demonstrated in Fig. 2.3 c). Increasing the magnetic field until a total flux of Φ_0 has entered the junction area leads to a phase difference between the ends of the JJ of 2π, which eventually means a complete suppression of any net current across the junction. This experiment is comparable to the single slit in optics, where the slit is the junction and the magnetic field corresponds to the viewing angle. Continuing this analogy, we know from optics that the transmission $I(\theta)$ is the Fourier transformation of the slit's spatial transmission $I(x)$ and follow for the transmission function of the junction $I_{\mathrm{c,eff}}(\Phi/\Phi_0)$ that it is the Fourier transformation of the junction's transmission $j_{\mathrm{c}}(x)$. This relation

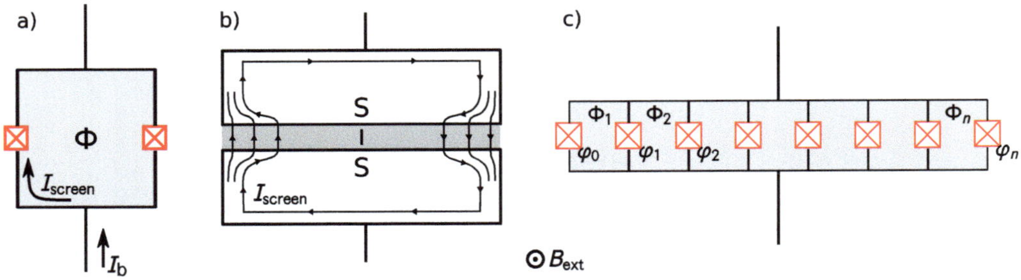

Figure 2.3: SQUID and JJ in a magnetic field. a) A DC SQUID consists typically of two JJs in parallel, where the flux Φ enclosed in the loop (gray area) induces a screening current I_{screen} and hence tunes the effective critical current for an additional bias current I_b across the SQUID. b) Cross-section through a single JJ with an insulating barrier (I, shaded) between two superconductors (S, white). In an external magnetic field, a screening current flows, but the penetration depth λ_L is larger in the insulating area, leading to a non-uniform current distribution. c) We can model a single junction by a parallel circuit of $n+1$ individual JJs, forming n SQUID loops. The phase drop across each junction is $\varphi_n = \varphi_{n-1} + 2\pi + \Phi_n/\Phi_0$. This means for $\sum_i \Phi_i = \Phi_0$ we get $\varphi_n - \varphi_0 = 2\pi$ and we can find for each junction with φ_i a junction with $\varphi_i + \pi$, whose currents cancel out each other, leading to a vanishing net current. If the field is further increased, this blockade is partially lifted, resulting in a Fraunhofer pattern for $I_c(B)$.

can also be derived in analytically exact ways and was demonstrated experimentally [93]. For the experiments presented in later chapters, we usually assume a rectangular shaped junction, leading to a sinc shaped critical current of

$$I_c\left(\frac{\Phi}{\Phi_0}\right) = I_c^0 \left| \frac{\sin \pi \frac{\Phi}{\Phi_0}}{\pi \frac{\Phi}{\Phi_0}} \right| = I_c^0 \operatorname{sinc} \frac{\Phi}{\Phi_0}.$$

2.5 The transmon qubit

Combining the findings that a qubit does not exactly have to be a two level system, but can also be an anharmonic multi-level system (Sec. 2.2) and that a superconducting Josephson junction offers a current-dependent inductor (Sec. 2.4), we intuitively find one of the most accessible realizations of superconducting qubits: the transmon qubit [17]. Although historically, the transmon qubit was derived as an improvement to the Cooper pair box or charge qubit, we follow here the more intuitive way of understanding the transmon as a nonlinear quantum oscillator. The most simplified electromagnetic oscillator consists of a capacitor and an inductor being connected to each other. The parabolic potential of this harmonic oscillator gives equidistantly spaced energy levels which cannot be used as a quantum bit. If we now replace the conventional inductor by a JJ, we find an anharmonic potential where the resonance frequency depends on the current flowing through the junction and therefore on the excitation number of the system.

For a more mathematical description, we define the degrees of freedom with which we want to describe the system. Instead of voltage and current, which are used in classical electrical engineering, we choose flux and charge here [94, 95]:

$$\Phi(t) = \int_{-\infty}^{t} V(t')\mathrm{d}t' \qquad \text{and} \qquad Q(t) = \int_{-\infty}^{t} I(t')\mathrm{d}t'.$$

We assume that the system is at rest in the beginning, $\Phi(-\infty) = Q(-\infty) = 0$. With Eq. (2.22) we directly see

$$\varphi = 2\pi\frac{\Phi}{\Phi_0}.$$

Expressed in these variables, the energy stored in the capacitor is now

$$W_\mathrm{C} = \frac{Q^2}{2C} \tag{2.24}$$

and the energy in the Josephson junction

$$W_\mathrm{JJ} = \int L(I)I\mathrm{d}I = \frac{\Phi_0}{2\pi}\int I(\varphi)\mathrm{d}\varphi = -I_\mathrm{c}\frac{\Phi_0}{2\pi}\cos\frac{2\pi\Phi}{\Phi_0}. \tag{2.25}$$

This gives the transmon Hamiltonian as

$$\begin{aligned} H &= \frac{Q^2}{2C} - I_\mathrm{c}\frac{\Phi_0}{2\pi}\cos\frac{2\pi\Phi}{\Phi_0} \\ &= 4E_\mathrm{C}\hat{N}^2 - E_\mathrm{J}\cos\varphi \end{aligned} \tag{2.26}$$

where we introduced the difference in number of Cooper pairs on the islands $\hat{N} = Q/2e$ and defined the charge energy $E_\mathrm{C} = e^2/2C$ and the Josephson energy $E_\mathrm{J} = I_\mathrm{c}\Phi_0/2\pi$. From the fact that $\hat{N}$ appears in this equation we see the transmon's close relation to the Cooper pair box, meaning that offset charges can still influence its eigenenergies. However, when increasing the capacitance, the charging energy E_C goes down and with that the contribution of the offset charges.

To describe the transmon as an anharmonic oscillator, we expand the cosine up to fourth order in φ, resulting in

$$H_\mathrm{approx} = 4E_\mathrm{C}\hat{N}^2 - E_\mathrm{J} + \frac{E_\mathrm{J}}{2}\varphi^2 - \frac{E_\mathrm{J}}{24}\varphi^4. \tag{2.27}$$

Recalling that $[\varphi, N] = \mathrm{i}$, we can compare the quadratic terms with N^2, φ^2 to a standard quantum harmonic oscillator, and express the operators as

$$\begin{aligned} N &= \frac{\mathrm{i}}{2}\left(\frac{E_\mathrm{J}}{2E_\mathrm{C}}\right)^{1/4}\left(\hat{b}^\dagger - \hat{b}\right) \\ \varphi &= \left(\frac{2E_\mathrm{C}}{E_\mathrm{J}}\right)^{1/4}\left(\hat{b}^\dagger + \hat{b}\right), \end{aligned} \tag{2.28}$$

with $\hat{b}^\dagger$, $\hat{b}$ being the creation and annihilation operators. This results in the Hamiltonian

$$H_{\text{approx}} = \sqrt{8E_C E_J}\left(\hat{b}^\dagger\hat{b} + \frac{1}{2}\right) - E_J - \frac{E_C}{12}\left(\hat{b}^\dagger + \hat{b}\right)^4$$
$$\approx \sqrt{8E_C E_J}\,\hat{b}^\dagger\hat{b} - \frac{E_C}{2}\hat{b}^\dagger\hat{b}\left(\hat{b}^\dagger\hat{b} + 1\right) + \text{const}, \tag{2.29}$$

where in the lower line, all terms without pairs of $\hat{b}$ and $\hat{b}^\dagger$ have been neglected, as they do not contribute to the transition energies when treating the quartic term in perturbation theory [96].

From this we easily find the energy of the m-th excited state to be

$$E_m \approx \sqrt{8E_C E_J}\,m - \frac{E_C}{2}\left(m^2 + m\right) + \text{const}. \tag{2.30}$$

The transmon in this approximation hence has an anharmonicity of $\alpha_m = E_{m,m+1} - E_{m-1,m} = -E_C$ and a fundamental qubit transition frequency $\hbar\omega_{01} = \sqrt{8E_C E_J} - E_C$. Typical values of $E_C/h \sim 200\,\text{MHz}$ for the samples used in this thesis are enough to prevent the excitation of higher levels by the bandwidth of the pulses used in the experiments. We can therefore take the transmon as an effective two-level system and hence as a qubit.

We note here that the approximation in Eq. (2.27) is only valid if φ is small, meaning that we are in the transmon regime with $E_J \gg E_C$ and are only considering the lowest excited states. The full solution to the transmon Hamiltonian in Eq. (2.26) can be calculated by using Mathieu functions. However, as it is numerically easier to implement the diagonalization of the Hamiltonian instead of using Mathieu functions, we only refer here to Ref. [17] for a full discussion of this solution and will use other methods to describe the exact eigenstates and eigenenergies.

With a decreasing E_J/E_C ratio, the influence of $\hat{N}$ gets stronger which we can probe by inducing offset charges by means of a gate voltage $\hat{N} \to \hat{N} - n_g$. The variation of the transition energies $E_m(n_g)$ with respect to the gate charge is given in Fig. 2.4 a)-c), where a decreased sensitivity on offset charges with increasing E_J/E_C is apparent. A comparison between the exact solution and the approximated transition energies from Eq. (2.29) is demonstrated in Fig. 2.4 d).

In the experiments, we can excite higher levels of the transmon by subsequently driving the individual transitions, i.e., perform the transition $|0\rangle \to |1\rangle$ by applying an ω_{01} photon, then the $|1\rangle \to |2\rangle$ transition by the corresponding ω_{12} photon, and so on. Alternatively, we can directly drive the $|0\rangle \to |m\rangle$ transition by the simultaneous irradiation of m photons of frequency ω_{0m}/m and the help of $m-1$ intermediate virtual energy levels. Due to multi-photon processes, these transitions start to appear only at higher drive powers compared to the single photon processes. The dispersive shift of the readout resonator, i.e., the readout signal, increases for the higher excited states [97].

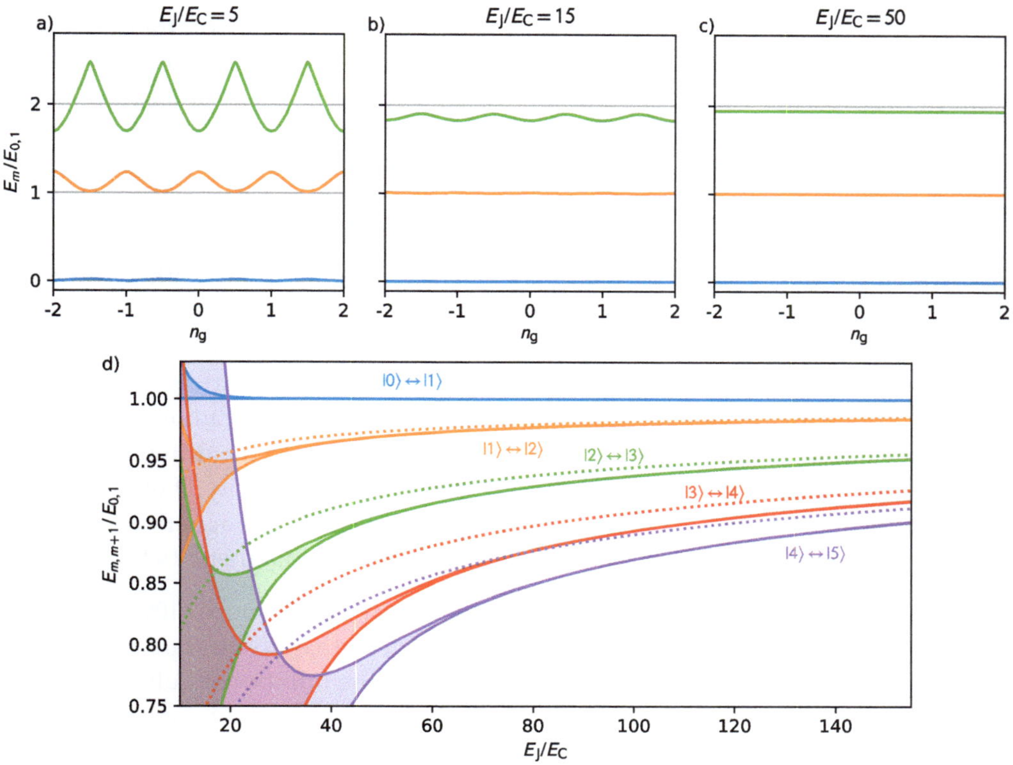

Figure 2.4: Transmon potential for different E_J/E_C ratios. a)-c) The normalized transmon energies for increasing ratios of E_J/E_C. With increasing capacitance and therefore decreasing E_C, the influence of single offset charges on the islands gets smaller and the energy levels get flatter. However, the level structure gets more harmonic, meaning that $E_{0,2}/E_{0,1} \approx 2$. d) To resolve these differences better, transition energies between adjacent levels are displayed. The solid lines and shaded areas highlight the range of transitions when varying the gate charge and therefore their spread shows the susceptibility against charge noise. Dashed lines show the approximation of Eq. (2.29) for comparison. It can be seen that this approximation is only applicable for the first transitions and deep in the transmon regime.

2.5.1 Concentric transmon

For a transmon qubit, basically only two ingredients are needed: a Josephson junction and a comparatively large shunting capacitance. Despite this simplicity, a variety of different designs has been proposed in the last decade and many improvements could be found. Starting from the first transmon design [17], depicted in Fig. 2.5 a), the capacitance was fabricated as interdigited finger structure, which enables a compact design but at the same time increases the electric field strength in the gap between the fingers. With this, the coupling to defects on the surface of the superconductor is increased, leading to higher losses and reduced coherence times.

This disadvantage could be evaded by the introduction of a microstrip transmon [98], where the electric fields are drawn into the low-loss substrate by increasing the spacing between the capacitor islands and introducing a metalized backside of the chip. Using inert TiN instead of the commonly used Al, the existence of surface defects could also be reduced. Besides the large size of this design, one drawback is the antenna-like structure which increases the qubit's radiative loss as well as spurious coupling of the qubit to other elements and signals on the chip. To counteract these disadvantages, the dipole moment was reduced by wrapping one electrode around the other, resulting in the concentric transmon [99], displayed in Fig. 2.5 c) and d). This design offers a reduced dipole moment and therefore reduced radiation as well as a confined field distribution. For this work, two different designs of the concentric transmon are used: one with a single JJ and therefore fixed frequency and one with a gradiometric design, featuring 2 JJs. As the gradiometric design is only sensitive to the difference in flux between the two inductance loops, it is – in contrast to the majority of other tunable superconducting qubits – ideally insensitive to homogeneous magnetic fields. This feature is of unique advantage for sensing applications in magnetic fields.

The concentric transmon was adopted and modified by other groups [100], offering a direct access to the qubit by moving the microwave feed lines to the third dimension. While this design might be beneficial for scaling up the number of qubits per chip, it reduces the flexibility and the ability to place other samples directly above the qubit chip, which we will later use in Ch. 7.

Besides the concentric transmon, other transmon designs exist and are commonly used. Here, in particular the "Xmon" design has to be mentioned, which offers multiple arms for capacitive coupling and therefore enables error correction schemes like the surface code [26, 101].

A detailed description of all samples used throughout this thesis and their characterization can be found in Sec. 4.3.

2.6 Loss mechanisms of superconducting qubits

The excitation of a qubit can decay into a variety of different channels and it is not always easy to identify the reasons leading to an experimentally observed relaxation time T_1. In the following, different decay mechanisms are listed, where the list makes no claim to be complete. Again, only qualitative argumentations are given, as the experiments later do not allow for a detailed differentiation. When several decay mechanisms are present, the resulting decay rate is the sum of the individual decay rates

$$\Gamma_{1,\Sigma} = \sum_i \Gamma_{1,i} = \sum_i T_{1,i}^{-1}. \tag{2.31}$$

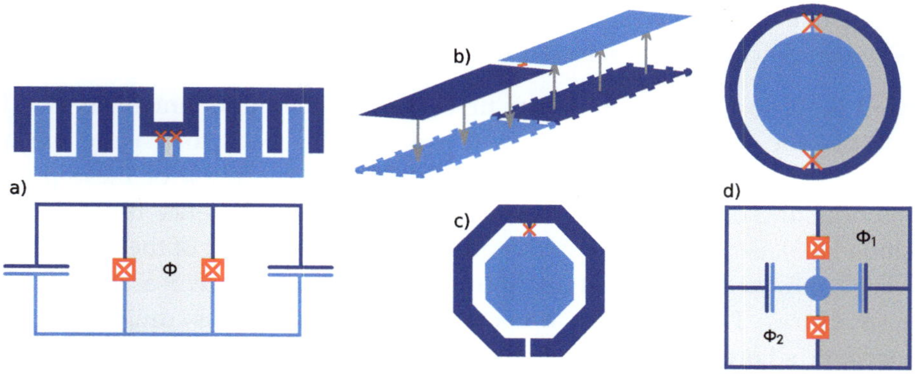

Figure 2.5: History of the concentric transmon design. a) The first transmon design features a SQUID loop (gray area) and interdigited finger capacitances [17]. b) Microstrip transmon with capacitive islands induces mirror charges in the ground plane (dashed frame) and concentrates the electric field lines into the substrate [98]. c) Concentric transmon with low dipole moment in the non-tunable version. d) Concentric transmon in gradiometric design. Only the flux difference $\Phi_2 - \Phi_1$ creates a current across the junctions and biases the qubit [99]. In a) and d), the lower graphs show an effective circuit diagram of the design with the JJs in red and flux-sensitive areas in gray. Figure adapted from [77].

2.6.1 Purcell and radiation loss

When the qubit is coupled to a harmonic resonator, we see from the $g\hat{\sigma}_-\hat{b}^\dagger$ term in the Jaynes-Cummings Hamiltonian Eq. (2.10) that the excitation can swap from the qubit to the resonator, where it can eventually decay, dissipate or leak into a third system. It is clear that besides the coupling strength g also the detuning $\Delta = \omega_\mathrm{q} - \omega_\mathrm{R}$ between the two systems and the bandwidth κ of the resonator influence the losses. For the bandwitdh, the full width at half maximum of a Lorentzian-shaped resonator is assumed. With this, we get [102]:

$$\Gamma_{1,\mathrm{purcell}} = 2g^2 \frac{\kappa/2}{(\kappa/2)^2 + \Delta^2}. \tag{2.32}$$

The Purcell loss can be estimated already at the design stage and is therefore usually optimized with respect to a good readout signal, given by the dispersive shift $\chi = g^2/\Delta$. The resonator bandwidth is limited by fabrication methods and materials.

This single-mode Purcell decay is however not only present for the designed readout resonator, but for every spurious mode, the qubit can couple to, e.g., modes of the sample box or other modes on the chip. Apart from that, the qubit excitation can also relax into non-resonant systems like transmission lines or other control lines by the multi-mode Purcell effect. Then, the Purcell loss is described by the impedance of the environment and the coupling capacity or inductance between qubit and environment

[102]. Due to the enhanced spurious coupling of a qubit with large dipole moment, these losses are commonly referred to as radiation losses.

To avoid these losses, one can either firstly design the environment to give little possibilities for spurious coupling, which eventually leads to placing the qubit inside a three dimensional cavity. Together with a very low κ of these polished superconducting cavities, this is the basis for the great coherence times of the 3D transmon [103]. Secondly, one could reduce the dipole moment of the qubit, leading to a very confined electric field distribution and therefore a decreased coupling to other structures. This approach was chosen for this thesis, as, unlike in the first approach, it is here possible to place other elements like wide-band transmission lines and lossy dielectrics close to the qubit.

2.6.2 Dielectric and TLS loss

As already mentioned, one goal for the design optimizations of the transmon was to concentrate the electric fields into the substrate, as the interfaces between superconductor and substrate can be controlled better than the surface of the superconductor to air, and the substrate itself is very pure. The contribution of the chip's substrate can usually be neglected: For intrinsic silicon, the loss tangent $\tan\delta = \mathrm{Im}(\varepsilon)/\mathrm{Re}(\varepsilon)$ is below $\delta < 10^{-7}$, where ε is the complex dielectric constant of the material. This would limit the qubit coherence to $\Gamma_1 = \delta\omega_{01} < (0.3\,\mathrm{ms})^{-1}$ [104, 105], and is not a limiting factor for state-of-the-art transmon qubits.

Since for the JJs of superconducting qubits the most common choice of material is $\mathrm{Al}/\mathrm{AlO_x}/\mathrm{Al}$, also the qubit's capacitance pads are often fabricated from Al for simplicity. One major problem is that aluminum oxidizes quickly at air or in wet processing steps, leading to the formation of amorphous aluminum oxide on the surface of the superconductor. In this amorphous material, mechanical degrees of freedom with an electric dipole moment exist, which hence couple to electric fields, absorb energy and increase the dielectric losses in the material. These so-called two level fluctuators (TLF) or in general two level systems (TLS) have been studied in great detail [34–36, 106] and are still subject to active research.

Despite the efforts taken to reduce the field strength in the gap of the capacitor and concentrate the field lines into the substrate, the loss tangent of $\delta_{\mathrm{AlOx}} = 3 \cdot 10^{-3}$ for amorphous aluminum oxide [105, 107] is so large that still the small portion of the electric field energy in the surface and interface oxides has to be taken into account. A more detailed study on an all-aluminum transmon [102] concludes that the contribution of the surface oxides is on the order of magnitude of current qubit coherence times and hence influences the measured T_1 values.

Although it is possible to fabricate the capacitance pads from a more inert material, the insulating barrier is exclusively fabricated from AlO_x and commonly, at least a short connection to the junctions is made from Al. Apart from that, also residues of processing materials, like resist or glue, or even water and bound oxygen molecules can lead to dielectric losses on the surface of the superconductor – independent of the material in use. Observing fluctuations in the qubit's coherence times, we can conclude that atomic fluctuators on the surface of the superconductor contribute considerably to the overall losses [31].

2.6.3 Quasiparticles

Quasiparticles (QPs) are excitations of the superconductor, e.g., split pairs, where one partner of the Cooper pair is missing. They can be described to be in the range from hole-like to electron-like, do not participate in the many-body wavefunction and therefore increase the energy of the system [86]. Hence, QPs cannot tunnel through the JJ without resistance and a QP current is in principal dissipative [93]. We note at this point that this explanation is very simplified, and in a more sophisticated analysis, the phase φ across the JJ has to be taken into account. In the transmon operation regime however, φ is small and we can simplify our analysis.

For a superconducting transmon qubit, the influence of QPs on the qubit's relaxation rate was in detail derived [108] and further calculated [109, 110] to be:

$$\Gamma_{\mathrm{QP}} \approx \frac{x_{\mathrm{QP}}}{\pi}\sqrt{2\omega_{01}\Delta_0/\hbar} \;\Rightarrow\; \Gamma_{\mathrm{QP}} \propto x_{\mathrm{QP}}.$$

The QP concentration x_{QP} is here normalized to the number of Cooper pairs and Δ_0 is the energy gap of the superconductor.

At finite temperatures, QPs are thermally excited, resulting in a QP concentration of $x_{\mathrm{QP,therm}}(T) = \sqrt{2\pi k_B T/\Delta_0}\exp\left(-\Delta_0/k_B T\right)$ [108]. For a temperature of $T \approx 20\,\mathrm{mK}$ and a thickness-dependent gap of thin aluminum films $\Delta_0 > 200\,\mu\mathrm{eV}$ [111], we would expect $x_{\mathrm{QP,therm}}(T) < 10^{-52}$, resulting in coherence times of thousands of years. In reality however, QP concentrations of $x_{\mathrm{QP}} \sim 10^{-7}$ are observed [103], giving $1/\Gamma_{\mathrm{QP}} \sim 1\,\mathrm{ms}$.

The origin of these excess QPs is still subject to research, and commonly, the following factors are considered: With the decreasing thermal conductivity at low temperatures, it becomes challenging to thermalize the qubit to the cryogenic environment and the chip temperature is usually assumed to be above the temperature of the base stage.

Also, any kind of radiation with $\hbar\omega > \Delta_0$ absorbed by the superconductor will break Cooper pairs and eventually create a cascade of QPs. To this end, filters for infrared light are typically included in the microwave wiring to the sample for high-coherence measurements. Impacts of high energy particles like cosmic radiation have also been observed to increase the QP density [112].

Apart from that, QPs can be excited by driving a strong current $I > I_c$ across the junction or applying a voltage. This for instance happens when strongly populating the cavity coupled to the qubit, which induces an oscillating voltage across the JJ, splitting Cooper pairs and hence creating QPs [113, 114].

Eventually, also magnetic fields lead the creation of QPs, where for parallel fields it holds $x_{QP} \sim B_{\parallel}^2$ [115] and the QP recombination rate, i.e., the relaxation rate of out-of-equilibrium QPs goes back with increasing magnetic fields [116].

Apart from shielding the samples against unwanted radiation and operating at the correct power levels, QP traps have been introduced, where a combination of materials with different gap energies are used such that the gap of the trap is lower than the one of the device, $\Delta_{0,D} > \Delta_{0,T}$ [117, 118]. From the fact that any excitation of the condensate needs to be above the condensation energy, it follows that $E_{QP} > \Delta_0$. Coming from material 1 to 2, the QP will relax, e.g. by interaction with phonons, until $\Delta_{0,D} > E_{QP} > \Delta_{0,T}$. Due to its now decreased energy, the QP cannot go back to material 1 and is trapped. With the reduced gap however, it is easier to excite QPs in the trap, for which reason the traps should be located with appropriate spacing from the JJ.

In the extreme case of $\Delta_{0,T} = 0$, a normal metal is used as trapping material [119–121]. This can also be achieved by suppressing superconductivity locally, e.g. by normal conducting flux vortices in the superconductor [113, 122, 123]. It has been demonstrated that the coherence of a QP limited fluxonium qubit can be enhanced by applying a small field, which in turn creates flux vortices at intended locations far from the junctions.

2.6.4 Movement of flux vortices

For type II superconductors, it was experimentally observed that their RF dissipation increases abruptly when $B_{c,1}$ is reached [124], leading to the conclusion that flux vortices contribute to the losses. In a gedankenexperiment we apply $B_z > B_{c,1}$ to a type II superconductor to create the vortex state. Due to the mutual repulsion of the vortices, they arrange in a regular hexagonal lattice in the equilibrium state to have the largest possible spacing. If we now apply a current along the x direction, we immediately get from Ampère's circuital law $\vec{\nabla} \times \vec{B} = \mu_0 \vec{j}_s$

$$\frac{\mathrm{d}}{\mathrm{d}y} B_z = \mu_0 j_x,$$

which means that the flux density increases in the y direction, leading to a gradient in the density of flux vortices. This means the vortices get pushed to positive values of y, or in other words, a Lorentz force acts on the vortices and they start moving.

As described in Eq. (2.16), on a contour around the flux vortex, the phase changes by 2π. Moving flux vortices therefore lead to a change in the phase of the surrounding superconductor and therefore via Eq. (2.22) to a voltage. The electric field corresponding to this internal voltage is orthogonal to the movement of the flux vortices and orthogonal to the magnetic field, hence it points along the direction of $\vec{j}$. With $\vec{E} \parallel \vec{j}$ however, dissipative processes take place and heat is generated.

We can come to the same conclusion from a microsopic analysis: Looking at a moving flux vortex, we see that it destroys Cooper pairs in front of it, as the inner of the vortex is normal conductive and QPs recombine behind the vortex, resulting in Cooper pairs. For the non-stationary case, we remember that the flux vortices move towards areas with higher B, meaning that the magnetic field is higher in the front of the vortex compared to its back. This means that splitting the Cooper pairs in the higher field takes less energy as will bee released by the condensation in the lower field area, leading again to dissipation. [125]

As a simple conclusion we find that moving flux vortices contribute to overall losses for RF applications [126], which was experimentally proved several times [127–129].

To reduce this dissipative influence of the vortex motion, a very common method is the creation of pinning centers, where the flux vortices preferably locate. These pinning centers can be created by structural defects or the doping of other materials, but too small pinning centers still allow the finite-sized vortex to wiggle around, leaving a contribution to the overall losses. Alternatively, for thin film superconducting structures, flux trap holes are structured into the design of the chip, where their size is smaller than the structures they should protect and corresponds to one flux quantum at the expected field strength [129–131].

When flux vortices are present in the superconductor and the field is ramped down again, flux vortices are not directly destroyed. Instead, vortices with opposing orientation start to enter the superconductor from the sides and only in an interface region between up and down vortices, the annihilate. This description was introduced by C. Bean and is hence called the Bean model [132, 133]. The mesoscopic magnetization of the superconductor is hence hysteretic and the configuration of flux vortices depends on the magnetic history of the sample. To decrease the amount of flux vortices in the superconductor, it is therefore not only sufficient to decrease the magnetic field or apply a counter field, but a demagnetization sequence has to be applied, where the magnetic field is slowly ramped between positive and negative values with decreasing amplitude [131]. Due to the multiple slow ramping processes, it is usually faster to switch off the field, warm up the sample above T_c and cool it back down in zero-field.

A good overview over magnetization effects and hysteresis of superconductors is also provided by Ref. [134].

3 Microwave theory

Like a large number of quantum systems, superconducting quantum bits have excitation energies in the range of several μeV to few meV. The transition frequencies of interest for coupling and sensing experiments are hence in the microwave regime. This means we can use commercially available microwave technology for our measurements and are able to describe the systems with well-established formalisms of microwave theory.

In this chapter I shortly recapitulate the theoretical description of microwave systems with a focus on the means and measures we will later use to characterize our systems. The chapter starts with general microwave networks and leads via a single resonator to two coupled oscillating systems which form hybridized modes.

3.1 Scattering parameters for a microwave network

To specify a network of microwave components with N ports, one of the most intuitive descriptions from an experimental point of view is the description by an $N \times N$ scattering matrix. The elements of this matrix correspond to the ratio of measured voltage at output port i to input voltage at port j:

$$\hat{S}_{ij} = \frac{V_i^{\text{out}}}{V_j^{\text{in}}} \text{ with } V_{k \neq j}^{\text{in}} = 0,$$

where the restriction means that we only drive the system at port j. For V_k^{in}, V_k^{out} the complex voltage amplitudes of the oscillating signals are taken, where the angle between the signals, $\arg(S_{ij}) = \arg(V_i^{\text{out}}/V_j^{\text{in}})$, is the phase difference between incoming and outgoing signal. Writing the incoming and outgoing amplitudes as vectors, we can write

$$\vec{V}^{\text{ out}} = \hat{S}\vec{V}^{\text{ in}},$$

or for the most commonly used case of a two-port network

$$\begin{pmatrix} V_1^{\text{out}} \\ V_2^{\text{out}} \end{pmatrix} = \begin{pmatrix} S_{11} & S_{12} \\ S_{21} & S_{22} \end{pmatrix} \begin{pmatrix} V_1^{\text{in}} \\ V_2^{\text{in}} \end{pmatrix}. \tag{3.1}$$

If the network under test does not contain any systems that can break time reversal symmetry, e.g., ferrites or active components, the network is called reciprocal, resulting in a symmetric scattering matrix $\hat{S}^{\mathrm{T}} = \hat{S}$. This means that exchanging ports 1 and 2 does not alter the behavior of the system [135].

Typically, the parameters of the scattering matrix have a characteristic dependence on the frequency. For active components and in the high power regime, the $\hat{S}$ parameters also become power dependent, resulting in $\hat{S}(\omega, P^{\mathrm{in}})$. The parameters of this 2×2 matrix can be experimentally acquired with a two port vector network analyzer (VNA) which allows for frequency sweeps at variable input powers. If the network has more than two ports, the full matrix can be measured by terminating all other ports with a matched load to avoid reflections and repeating this procedure for all combinations.

Due to the high dynamic range of the parameters, they are typically given in logarithmic units, standardized as decibel. One Bel (i.e., ten decibel) corresponds to a ratio of ten in powers, or in general

$$L_P = 10 \log_{10} \frac{P_1}{P_{\mathrm{ref}}}.$$

Two points in this definition are commonly disregarded, which can lead to errors, misunderstandings or simply imprecise phrasing: First, it is important to note that the unit Bel always represents a relation of powers, no matter which size of quantity is actually displayed. So if we want to refer to a voltage ratio in terms of dB, we have to consider the power at the characteristic impedance Z given by $P = V^2/Z$ and hence

$$L_P = 10 \log_{10} \frac{P}{P_{\mathrm{ref}}} = 10 \log_{10} \frac{V^2}{V_{\mathrm{ref}}^2} = 20 \log_{10} \frac{V}{V_{\mathrm{ref}}},$$

which means that a prefactor of 20 appears instead of 10 if we relate voltages, currents, amplitudes, and other quantities.

Second, the Bel always requires a reference value P_{ref}. For the $\hat{S}$ parameters, this is obviously the input power, but it is also possible to take a fixed reference value to state absolute power values. For the power levels used throughout this thesis, a very common unit is decibel-miliwatt (dBm), where the reference value is $P_{\mathrm{ref}} = 1\,\mathrm{mW}$. A power level of $L_P = -30\,\mathrm{dBm}$ then corresponds to an absolute power value of $P = 10^{L_P/10}\mathrm{mW} = 1\,\mu\mathrm{W}$.

Using logarithmic units becomes especially handy when handling attenuation and amplification factors of microwave components, as they can be simply added to the output power of the microwave source in dBm, as long as all factors are given in dB.

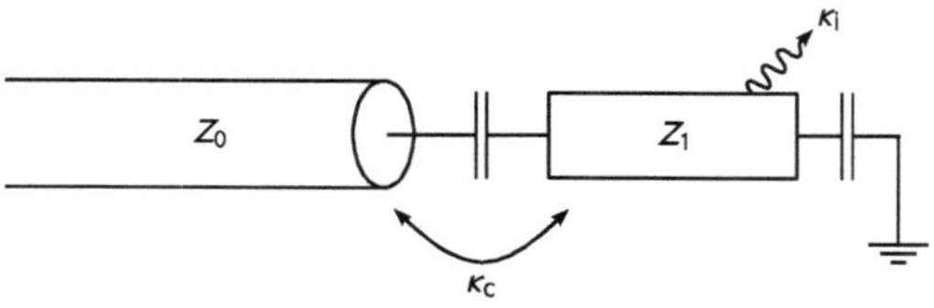

Figure 3.1: Schematic of a resonator. A transmission line with impedance Z_0 is capacitively coupled to a resonator with impedance Z_1. The interaction or energy exchange rate between the systems is given by κ_c. Inside the resonator, losses to other systems and dissipation can occur, represented by a rate κ_i.

3.2 Microwave resonators

When the characteristic impedance of a microwave transmission line Z_0 changes at one point abruptly to a value of Z_1, a fraction of the wave is reflected at this point, with [135]

$$S_{11} = \frac{Z_1 - Z_0}{Z_1 + Z_0}, \tag{3.2}$$

where we note that for the case $Z_1 < Z_0$ a phase difference of π can be observed. This scenario is depicted in Fig. 3.1. If we now introduce a second change in impedance at a distance l form the first change, the wave is reflected at both impedance mismatches and we can observe standing waves at a frequency of

$$\frac{\omega_n}{2\pi} = \frac{v_p}{\lambda_n} = \frac{n}{2l}\frac{c}{\sqrt{\varepsilon_{eff}}}$$

for mode number n, where we assumed that we have a $\lambda/2$ resonator, i.e., the impedance outside the resonator is on both sides higher (or on both sides lower) than the impedance inside the resonator. Here we used the phase velocity $v_p = c/\sqrt{\varepsilon_{eff}}$ in the medium, depending on the vacuum speed of light c and the effective dielectric constant in the medium ε_{eff}. The mode-number dependent wavelength is given by $\lambda_n = 2l/n$, with $n \in \mathbb{N}_{>0}$.

A resonator can also be created by connecting a capacitor C with an inductance L, where a resistor R is typically taken into account to sum up the losses. If the resistive losses are low, the oscillation frequency is given by $\omega_0 = 1/\sqrt{LC}$. Note that this resonator ideally has no higher harmonics, whereas the transmission line resonator has an infinite number of modes $\omega_n = n\omega_1$. The following discussions however apply to both types of resonators and L, C, and R denote the corresponding inductance, capacitance and resistivity of an equivalent lumped-element circuit [135].

3.3 Quality factors

Besides the resonance frequency of a harmonic oscillator, an important quantity is its quality factor Q, describing the ratio between the total energy stored in the oscillator E_{tot} and the energy dissipated per cycle $E_{\mathrm{diss}} = P_{\mathrm{loss}}/\omega_0$:

$$Q = \omega_0 \frac{E_{\mathrm{tot}}}{P_{\mathrm{loss}}} = \frac{\omega_0}{\kappa},$$

where κ is the full width at half maximum (FWHM) of a Lorentzian shaped frequency response and corresponds to the inverse photon lifetime [135]. In other words, the quality factor describes π times the number of oscillations the resonator can perform, until its amplitude has dropped to $1/\mathrm{e}$.

Photons in a resonator can be lost either by coupling into a different system, i.e., the transmission line, or by internal losses due to ohmic dissipation and dielectric losses. The quality factor is therefore divided into a coupling and an internal quality factor, Q_{c} and Q_{I}:

$$\begin{aligned} \frac{1}{Q_{\mathrm{L}}} &= \frac{1}{Q_{\mathrm{i}}} + \frac{1}{Q_{\mathrm{c}}} \\ \text{or } \kappa_{\mathrm{L}} &= \kappa_{\mathrm{i}} + \kappa_{\mathrm{c}}, \end{aligned}$$

where the loaded quality factor Q_{L} accounts for the total losses.

While Q_{c} in our case is mainly given by geometric values and therefore fixed for a given design, the internal quality factor for a superconducting resonator depends on a more sophisticated set of variables. Amongst others, these are the choice of materials, details of the fabrication process and the interface cleanliness. It can also be altered in an experiment, e.g., by applying a magnetic field. Here, the same mechanisms as previously discussed for the superconducting qubit apply, in particular losses due to quasiparticle excitations (Sec. 2.6.3) and flux vortex movement (Sec. 2.6.4).

3.4 Reflection spectrum of a resonator

In the experiments, we observe a transmission line resonator whose one end is capacitively coupled to a transmission line of impedance Z_0, while the other end is open. If we assume that the losses are generated by the oscillating current in the resonator, we can assume the form of a serial RLC circuit, which has an input impedance of [135]

$$Z_1(\omega) = Z_0 \left(\frac{Q_{\mathrm{c}}}{Q_{\mathrm{i}}} + 2\mathrm{i} \frac{\omega - \omega_0}{\omega_0} Q_{\mathrm{c}} \right),$$

close to its resonance frequency. Plugging this impedance into the scattering parameter from Eq. (3.2), we get

$$S_{11} = \frac{Z_1 - Z_0}{Z_1 + Z_0} = 1 - \frac{2Q_\mathrm{L}/Q_\mathrm{c}}{1 + 2\mathrm{i}Q_\mathrm{L}\frac{\Delta\omega}{\omega_0}}, \tag{3.3}$$

where we introduced the detuning $\Delta\omega = \omega - \omega_0$. From this formula we see that for large detuning, i.e., in the off-resonant case, we always are close to $S_{11}^{\mathrm{OR}} = 1$. At resonance however with $\Delta\omega = 0$, we get $S_{11}^{\mathrm{R}} = 1 - 2Q_\mathrm{L}/Q_\mathrm{c}$, being again real. If we shift the result of Eq. (3.3) by the mean of $S_1^{\mathrm{OR}}1$ and $S_1^{\mathrm{OR}}1$, we get

$$\tilde{S}_{11} = (S_{11}^{\mathrm{OR}} + S_{11}^{\mathrm{R}})/2 - S_{11} = \frac{Q_\mathrm{L}}{Q_\mathrm{c}}\left(-1 + \frac{2}{\delta^2+1} - 2\mathrm{i}\frac{\delta}{\delta^2+1}\right)$$

with $\delta = 2Q_\mathrm{L}\Delta\omega/\omega_0$. This gives

$$\begin{aligned} \left|\tilde{S}_{11}\right| &= Q_\mathrm{L}/Q_\mathrm{c} \\ \arg(\tilde{S}_{11}) &= \mathrm{arctan}_2(-2\delta, 1-\delta^2) = 2\arctan\frac{2Q_\mathrm{L}\Delta\omega}{\omega_0}, \end{aligned}$$

where $\mathrm{arctan}_2(y,x)$ is the function providing the complex argument to $z = x + \mathrm{i}y$. From this result we see that Eq. (3.3) represents a full circle in the complex plane, centered around $x = 1 - Q_\mathrm{L}/Q_\mathrm{c}$. For the case $Q_\mathrm{L} > Q_\mathrm{c}/2$, and therefore $Q_\mathrm{i} > Q_\mathrm{c}$, the origin of the complex plane is inside the circle and the phase of S_{11} rolls off over 2π. Since more energy is lost by the coupling than by internal dissipation, this is called the over-coupled case. For the under-coupled case, with $Q_\mathrm{i} < Q_\mathrm{c}$, most energy is dissipated internally, the origin is outside the circle, and the phase only changes by less than π. The case of $Q_\mathrm{i} = Q_\mathrm{c}$ is called critical coupling. This behavior is demonstrated in Fig. 3.2.

We also see that we can access the ratio $Q_\mathrm{L}/Q_\mathrm{c}$, and with that the individual values of Q_i and Q_c when fitting a circle to the complex scattering data. To this end, we utilize the circle fit routine [136], where more parameters are introduced to correct for imperfections. Now, Eq. (3.3) reads

$$S_{11} = A\,\mathrm{e}^{\mathrm{i}\alpha}\,\mathrm{e}^{-\mathrm{i}\omega\tau}\left(1 - \frac{2Q_\mathrm{L}/Q_\mathrm{c}}{1 + 2\mathrm{i}Q_\mathrm{L}\frac{\Delta\omega}{\omega_0}}\mathrm{e}^{\mathrm{i}\phi}\right), \tag{3.4}$$

where A is the maximum amplitude of the signal and accounts for overall losses and α corresponds to an overall phase offset. τ denotes the time the signal takes to travel through the setup. This means $v_\mathrm{p}\tau$ is the effective length of all wiring, which gives a frequency dependent phase offset. Finally, ϕ accounts for impedance mismatches and gives a rotation of the circle around the off-resonance point, which can be seen by a skewed shape of the amplitude signal. Measured data fitted by this equation is presented in Fig. 3.3.

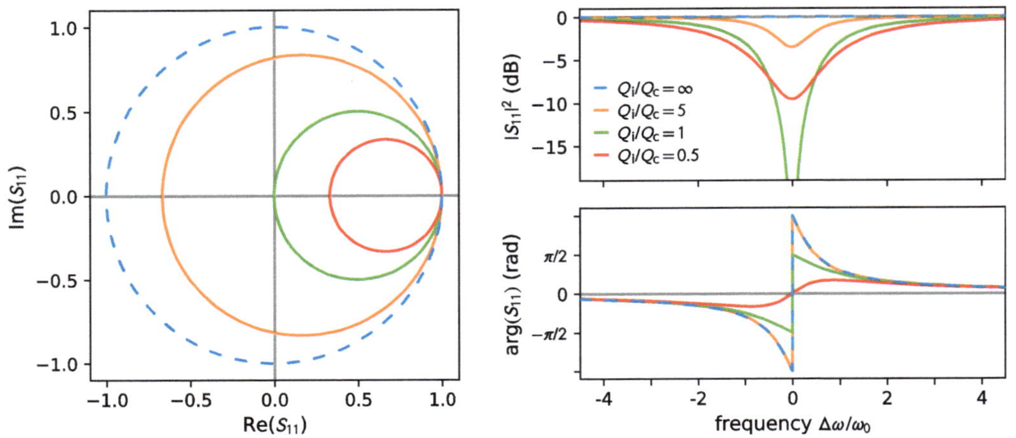

Figure 3.2: Visualization of S_{11} for different Q_i. In the complex plane (left) all calculated traces for $S_{11}(\Delta\omega/\omega_0)$ are on a circle starting at the off-resonance point $S_{11} = +1$. For an internal quality factor much higher than Q_c, the circle has a radius of 1 and no amplitude signal is visible (top, dashed blue line). In the phase signal (bottom), a 2π rolloff is visible. The greatest change in the amplitude signal is visible for a critically coupled resonator with $Q_i = Q_c$ (green line). Measurement data can be found in Fig. 3.3.

3.5 Average photon number

For the observation of quantum phenomena and for the investigation of power dependent effects, it is necessary to know the average number of photons $\langle n \rangle$ in the resonator. This is related to the energy stored in the resonator by $E_{tot} = \langle n \rangle \hbar\omega_0$, where we can use the definition of the quality factor in Eq. (3.3) to obtain

$$\langle n \rangle \hbar\omega_0 = P_{loss}\frac{Q}{\omega_0}. \tag{3.5}$$

In the steady state, the dissipated power equals the difference between incoming and outgoing power $P_{loss} = P_{in} - P_{out} = P_{in}(1 - |S_{11}|^2)$. Since the losses due to coupling are accounted for in P_{out}, the quality factor in Eq. (3.5) only has to cover internal losses. At resonance, we can take $S_{11}^R = 1 - 2Q_L/Q_c$, finally leading to

$$\langle n \rangle = \frac{Q_i P_{in}}{\hbar\omega_0^2}(1 - |S_{11}|^2) = 4\frac{Q_i P_{in}}{\hbar\omega_0^2}\left(\frac{Q_L}{Q_c} - \frac{Q_L^2}{Q_c^2}\right) = P_{in}\frac{4}{\hbar\omega_0^2}\frac{Q_L^2}{Q_c}. \tag{3.6}$$

For constant quality factors, the mean photon number in the cavity is hence proportional to the applied power $\langle n \rangle \propto P_{in}$, when excited on resonance.

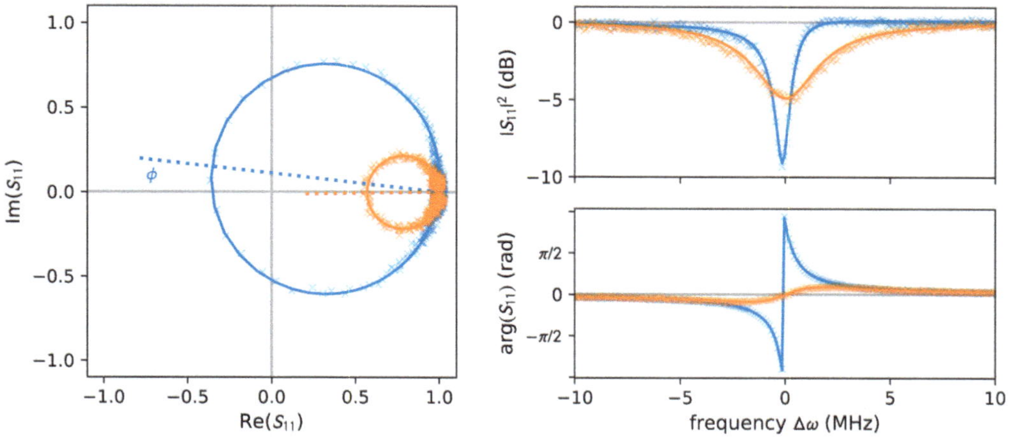

Figure 3.3: Measured data of over- and under-coupled resonators. A resonator with two different internal quality factors caused by different surrounding magnetic fields. The amplitude data are normalized prior to the fit to account for the transmission through the microwave setup. The solid lines represent the results of the circle fit, where the dashed lines show the "axis" of the circle, i.e., the connection between off resonance point and resonance. The measurement with high Q_i (blue) shows a well visible tilt of this axis, indicated by the angle ϕ in Eq. (3.4), caused by impedance mismatches. This can also be seen in the $|S_{11}|^2$ data as an asymmetric curve shape, where the right shoulder is higher than the left one. The quality factors extracted from the fit are $Q_i = 17\,300$, $Q_c = 8\,200$ for the resonator shown in blue and $Q_i = 2\,300$, $Q_c = 8\,100$ for the orange one.

3.6 Coupling resonant systems

If we not only take one microwave resonator but use two oscillating systems with $a_i(t) = a_i\, e^{i\omega_i t}$ and let them exchange energy at a rate g, their oscillation modes will hybridize when their resonance frequencies are close, $|\omega_1 - \omega_2| \sim g$, leading to new resonance frequencies. We can calculate these eigenfrequencies by solving

$$\begin{pmatrix} \omega - \omega_1 & g \\ g & \omega - \omega_2 \end{pmatrix} \begin{pmatrix} a_1 \\ a_2 \end{pmatrix} = 0.$$

This gives two solutions for ω:

$$\omega_\pm = \frac{\omega_1 + \omega_2}{2} \pm \frac{1}{2}\sqrt{(\omega_1 - \omega_2)^2 + 4g^2}, \tag{3.7}$$

which is commonly called avoided level crossing or anticrossing as the resonance frequencies "repel" each other. The two eigenfrequencies never approach closer than $\omega_+ - \omega_- \geq 2g$, where the equal sign holds for the resonant case, $\omega_1 = \omega_2$. To resolve the anticrossing experimentally, we need an external tuning parameter which changes the difference between the two bare eigenfrequencies. This behavior is depicted in Fig. 3.4.

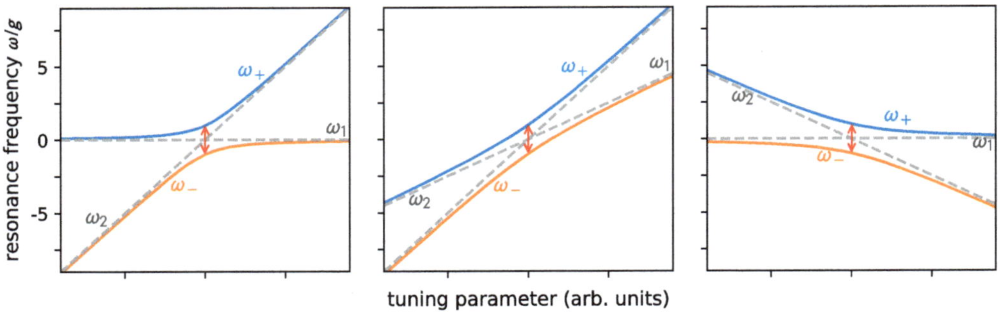

Figure 3.4: Avoided level crossings with constant coupling strength. Solution of Eq. (3.7) for a constant coupling factor $2g$ (red arrow) and different behavior of the bare eigenfrequencies (gray dashed lines) to the tuning parameter. We see that from the experimentally measured data we can fit and calculate back to the bare frequencies, the coupling and the dependence of the bare frequencies on the tuning parameter.

While Eq. (3.7) only gives the resonance frequencies, we do not obtain an expression for the linewidths of the resonances depending on the frequency. For a full expression of the S_{11} parameters, we could use the input-output formalism [137], which would require a more in-depth introduction here. Instead, we follow a more argumentative deduction here, based on the previous findings, which comes to the same result.

For the following argumentation, it is useful to rewrite the reflection coefficient of a single resonator from Eq. (3.3) using the linewidth κ instead of the quality factors. We hence find

$$S_{11} = 1 - \frac{2\kappa_c}{\kappa_L + 2i\Delta\omega_1}. \tag{3.8}$$

When we now add a second system, we have to account for the losses to this system, represented by κ_2, by increasing the loaded linewidth $\tilde{\kappa}_L = \kappa_L + \kappa_2$. The amplitude not reflected by the second resonator is given by $1 - S_{22} = 4g/(\kappa_{i,2} + 2i\Delta\omega_2)$, where $\kappa_{i,2}$ is the internal linewidth of the second resonator. The additional factor of 2 comes from the fact that g represents the half width at half maximum in contrast to κ. This quantity is mediated by the coupling with $\kappa_2 = g(1 - S_{22})$ so that Eq. (3.8) results in

$$S_{11} = 1 - \frac{2\kappa_c}{\frac{4g^2}{\kappa_{i,2}+2i\Delta\omega_2} + \kappa_L + 2i\Delta\omega_1}. \tag{3.9}$$

This very hand-wavy argumentation should not be understood as a full and exact derivation but rather as an explanation for the result, which is exactly derived by input-output theory in Ref. [64].[1]

[1] We note again that κ in this work is always understood as full width at half maximum linewidth (FWHM). In the cited reference, the half width at half maximum linewidth is used instead.

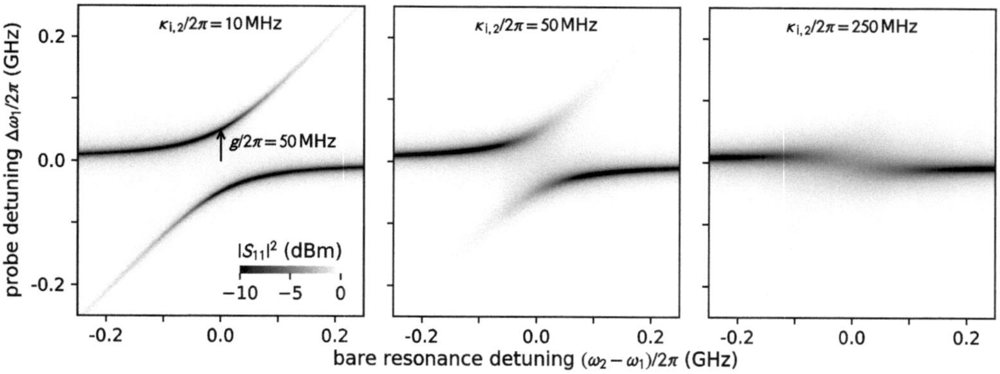

Figure 3.5: Spectrum of $|S_{11}|^2$ for different internal linewidths. Calculation of avoided crossings according to Eq. (3.9). The resonance frequency of the first resonator, which couples to the transmission line, is fixed and the detuning of the second resonator is given on the x axis. The coupling of $g/2\pi = 50\,\mathrm{MHz}$ is the same for all images. The linewidths of the first resonator are chosen to be $\kappa_c/2\pi = 10\,\mathrm{MHz} = 2\kappa_{i,1}/2\pi$, giving an overcoupled resonator and therefore a good visibility in the amplitude data. The internal linewidth of the second resonator is chosen as $\kappa_{i,2} = g/5$ (left), $\kappa_{i,2} = g$ (center) and $\kappa_{i,2} = 5g$ (right) to visualize the transition from strong to weak coupling.

If we calculate the $|S_{11}|$ spectrum for different detunings $\Delta\omega_1$, $\Delta\omega_2$ and vary the participating linewidths, we see a transition of the frequency response depending on $g \gtrless \kappa_{max}$, where κ_{max} is the largest of the included linewidths. For the variation of $\kappa_{i,2}$, this is depicted in Fig. 3.5. In the case $\kappa_{i,2} < g$ (left), the avoided crossing is well visible, but it blurs out when increasing the losses of the second system. For $\kappa_{i,2} > g$ (right), only a weakening of the first resonator's resonance line can be observed when the systems are in resonance. In this limit, the excitation in the second system decays faster than it can swap back into the first system. In other words, the systems cannot couple resonantly but the second resonator is just another loss channel for the first one. This also explains the line broadening of the first resonator when tuned to the same frequency. As the anticrossing is only visible for $g > \kappa_{max}$, we call this regime the strong coupling regime.

It is important to note here that the solution of Eq. (3.7) is not affected by these changes of κ_{max} and always represents the energies of the avoided crossing, meaning that Eq. (3.7) can only be taken into account for the strong coupling regime.

The previous considerations only studied steady-state solutions when exciting one system stronger than the other one, such that they continuously exchange energy. Then, a beating, .i.e., a repulsion of the resonance frequencies as shown in Eq. (3.7), is observed. If the two systems are equally excited by microwave pulses, they oscillate in phase (or anti-phase) and no energy is exchanged. We studied this topic in-depth by pulsed measurements on a microwave cavity coupled to a magnetic oscillator and published the results in Ref. [138].

4 Experimental methods

In this chapter, I give an overview of the experimental methods used throughout this thesis. I start with the cryogenic environment, which includes microwave and DC wiring and then go to the microwave measurement setups, which enable continuous wave and pulsed measurements. A remark on the home-grown software package QKIT concludes the measurement part, after which the qubit samples are presented. The chapter ends with a short summary of relevant fabrication methods.

4.1 Cryogenic setup

To set the ambient conditions necessary for the operation of superconducting quantum bits, we need to cool the quantum chip down to several tens of millikelvin and provide microwave and DC connections for control and readout. Such low temperatures are not only necessary to reach the superconducting state, but also to reduce the thermal population of the qubit and even more important the quasiparticle density in the superconductor, which would otherwise strongly limit the qubit coherence.

4.1.1 Dry dilution refrigerators

To reach the regime of about 20 mK in continuous operation, the choice of cooling systems falls almost exclusively on dilution refrigerators. These cryostats operate with a mixture of ^{3}He and ^{4}He, where the two isotopes differ in their behavior at low temperatures. Below 2.2 K, the bosonic ^{4}He atoms can fall into a condensate state, which is known as superfluidity. ^{3}He atoms however are of fermionic nature and do not[1] reach a condensate state, but stay in the liquid state with normal viscosity. Below this transition temperature, the two isotopes undergo a phase

[1] Actually, it was found that ^{3}He reaches a superfluid phase at 2.5 mK by the formation of Cooper pairs [139]. This is however not of practical use and not relevant here.

separation, one with a high and one with a low concentration of ^{3}He. Due to their different mass densities, they also separate spatially.

The small amount of ^{3}He atoms in the dilute phase can move in the superfluid liquid resistance free and therefore almost as in vacuum. The transition of ^{3}He atoms from the concentrated to the dilute phase therefore equals an evaporation process, where the transition energy is taken from the thermal energy of the environment. Since there is an equilibrium concentration of about 6 % of ^{3}He atoms in the dilute phase, this equilibrium would be quickly reached and the cooling power would vanish. Therefore, the ^{3}He concentration has to be artificially decreased to push the system away from equilibrium and get a continuous cooling.

This is done by connecting the dilute phase to a still chamber, which can be heated and where a turbomolecular pump is connected. Due to their different partial pressure, mainly ^{3}He evaporates in the still, which is then cleaned in a liquid nitrogen coldtrap, compressed, and liquefied on its way to the dilution unit, where it is inserted into the ^{3}He-rich phase. A schematic of this is flow is given in Fig. 4.1.

Although this method allows in principle for arbitrarily low temperatures, heat input from the cryogenic circle and via mechanical connections limit the base temperature of bare ^{3}He/^{4}He dilution refrigerators to about 10 mK. With additional wiring for microwave and DC connections, this typically increases to a value of $10 - 20$ mK for multi-purpose cryostats.

The pre-cooling to reach the phase separation temperature can either be done by reservoirs of cryogenic liquids like nitrogen and ^{4}He, or by a closed cycle pulse tube refrigerator. The latter has the advantage that no infrastructural facilities for the liquefaction and storage of liquid nitrogen and helium are necessary and that the cryostat can be operated for weeks without user interaction, which is why it was used for this thesis. The pulse tube cooler acts analog to an inverted Stirling motor, where the working gas itself is used as a second piston, avoiding moving parts in the cold area. Requiring only electric power and cooling water, a two-stage pulse tube cooler can cool down to typically 3 K. To bridge the gap to the phase separation at 2.2 K, the ^{3}He/^{4}He mixture is compressed to 3 bar, precooled at the pulse tube stages and then expanded via a Joule-Thompson (JT) valve, leading to a sudden drop in temperature. In normal operation, the pressure is reduced to about 500 mbar and additional cooling power of the JT valve is not required.

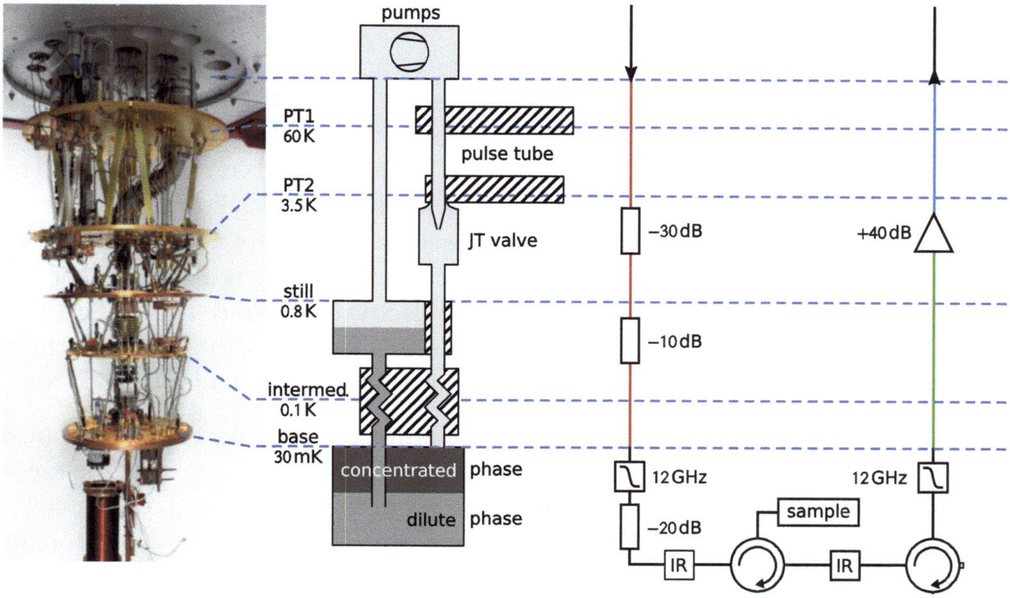

Figure 4.1: Cryogenic measurement setup. A full photograph of the Oxford cryostat is shown on the left, where the according temperature stages are marked. On the right, the cryogenic microwave setup is illustrated, where the colors of the lines correspond to the choice of material for the coaxial cables: stainless steel (red), CuNi (blue), superconductor (green), and regular copper coaxial cables (black).

In this thesis, experiments were carried out in two different cryostats of different manufacturers, one BlueFors LD 400 and one Oxford Triton 200. They differ in some engineering details and in their inner configuration but rely on the same physical principles for cooling. Great care was taken in all cryogenic setups to reduce the heat input on the temperature stages, since the minimum working temperature is limited by the cooling power. Therefore, any wiring between two stages was carefully selected in terms of material and wire cross section, to reduce the thermal conductivity. Also, Ohmic dissipation of current-carrying components has to be taken into account.

4.1.2 Cryogenic microwave setup

Having a large metallic cross-section, coaxial wires can contribute significantly to the heat input at low temperature stages. Therefore, each of the three components – outer conductor, dielectric, and inner conductor – needs special consideration. Being metallic, the inner and outer conductor are the most obvious crucial parts in cryogenic microwave cables. While regular cables are based on copper, we use more resistive alloys like copper-nickel or stainless steel in our cryostats to reduce the thermal conductance. This increases at the same time the electrical resistance and gives rise to microwave losses. The increased electrical resistance is beneficial as its basis is formed by an increased electron-phonon-interaction, which helps to thermalize the electrons.

Besides the electronic and the phononic temperature, we also take care of the noise temperature and power of the signal we send to the sample in the cryostat. The noise temperature of the input signal is approximately room temperature and commercially available electronics create signal strengths which are orders of magnitude stronger than what is needed for few photon operation in the cryostat. We therefore have to attenuate the signal, where it is obviously beneficial to put the attenuators at the cold stages of the cryostat, since every resistor creates Johnson noise corresponding to its temperature. If the signal is attenuated at N stages by an attenuation factor $a_i > 1$ at temperature T_i, the final noise temperature at the sample is given by

$$T_\mathrm{n} = T_1 + \sum_{i=1}^{N} \frac{T_{i+1}}{\prod_{j=1}^{i} a_j},$$

where $T_{N+1} = 300\,\mathrm{K}$ is the signal input at room temperature and T_1 the base temperature of the cryostat. Although it seems beneficial to pack all attenuation to the coldest stage in order to get a low noise temperature, this is experimentally not feasible since the attenuators would become very warm due to the dissipated signal and hence increase T_1. Typically, the total attenuation of around 60 dB is distributed amongst the three coldest stages and additional 10 dB are associated with the loss by the resistive coaxial wires.

For both fridges used in this thesis, the attenuation was distributed onto the base, still and second pulse tube stage at temperatures of about 30 mK, 800 mK and 3.5 K, respectively. For the BlueFors fridge, a combination of 20/10/30 dB was used and for the Oxford fridge 20/20/20 dB. This gives a noise temperature of $T_\mathrm{n,B} = 41\,\mathrm{mK}$ and $T_\mathrm{n,O} = 38\,\mathrm{mK}$.

While this high attenuation is intended for the input signal path, this scheme is not applicable to the output signal path, where the signal power is in the low photon regime. Even the attenuation of the resistive coaxial wires would lead to the loss of 9 out of 10 photons. Instead, we use a coaxial cable made of superconducting niobium which provides ideal electric conductance while at the same time having no thermal conductance. This wire is used from the base plate to the 4 K stage, where a HEMT[2] amplifies the signal by about 40 dB and stainless steel cables can be used afterwards.

The third component of coaxial wires is the dielectric, which is typically PTFE, also known as Teflon. Since PTFE is transparent in the infrared regime and infrared raditation from room temperature can create quasiparticles and therefore limit the coherence, some effort was spent in the field and in our group to develop filters for infrared radiation. While a significant improvement of coherence times was absent, some filters had mechanical problems leading to visible cracks and improper microwave transmission after a temperature cycle of the cryostat. The IR filters have therefore only been used in the BlueFors cryostat.

At the base stage of the cryostat, we use a circulator to be able to perform a reflection S_{11} measurement of our sample with our setup of dedicated input and output lines and a second one to shield the sample from the electronic noise coming from the HEMT. Additional band-pass filter are used to shield the sample from any unwanted microwave signals which could eventually pass the circulators out of their working band. Depending on the number of samples in the cooldown, also a 6-port microwave switch was used between circulator and sample, to be able to measure multiple samples without warming up and opening the fridge.

4.1.3 Magnetic field bias

Both fridges feature a superconducting solenoid, which can create a homogeneous magnetic field. In the BlueFors fridge, a commercially fabricated solenoid with 5 796 windings creates 47.5 mT/A and in the Oxford fridge, a self-made solenoid of 10 330 windings creates 81.5 mT/A.

In the BlueFors cryostat, the coil is mounted to the still stage such that the sample can be fixated to the base stage (see Fig. 4.2). This causes less heating on the base stage when ramping the magnetic field due to eddy currents and is beneficial in the case of a quench of the coil. We took great care in this case that no part of sample or wiring touch the solenoid body to avoid a thermal short. Also, the sample temperature is assumed to be slightly higher due to the influence of radiation from the solenoid. To this end, the coil in the Oxford fridge is mounted to the base stage and the ramping rate of the magnetic field was reduced here.

[2] HEMT: High electron mobility transistor

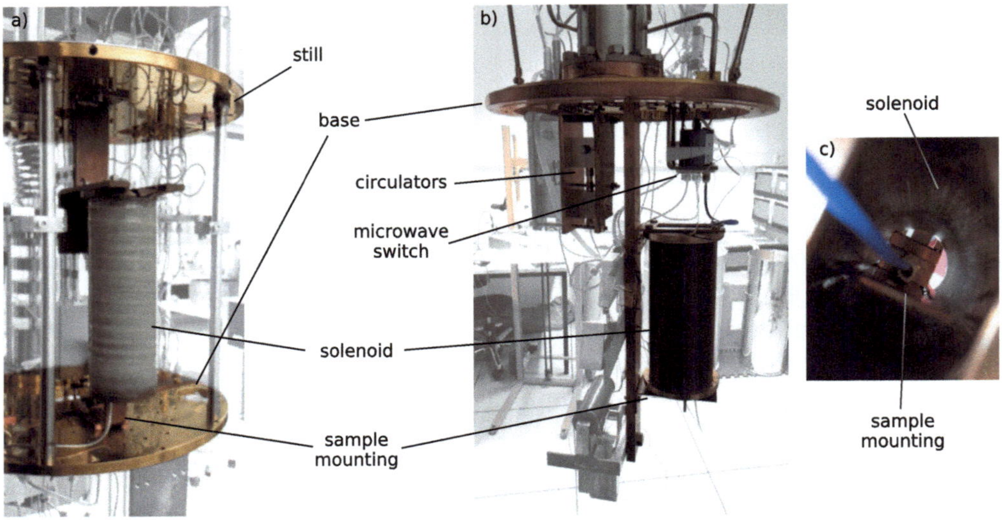

Figure 4.2: Solenoid configuration in both fridges. a) The solenoid in the BlueFors fridge is mounted to the still stage with the sample reaching in from the base stage below. b) In the Oxford fridge, both solenoid and sample are fixated at the base stage. c) The sample is aligned by eye such that the chip is at the very center of the diameter.

The samples in the solenoid are aligned by eye, which leaves a small probability for an off-center position and a tilt against the field lines. Although the pure qubit coherence as presented in Ch. 6 might be improved by a better alignment and the use of compensation coils, we believe that this is not a realistic scenario as soon as magnetic materials are introduced on top of the qubit, which will alter the magnetic field configuration.

To supply the comparatively high currents down to the base stage, different materials are used: Between room temperature and the 70 K stage, massive copper wires are taken which increase the thermal load on this stage. For the lower cooling power of the 3.5 K stage, this is not feasible and ribbons of high temperature superconductor on a stainless steel matrix are used here. Below, a NbTi superconducting wire in a Cu or CuNi matrix is used. The technical realization of these connections is in detail described in the PhD thesis of my colleague Marco Pfirrmann [140].

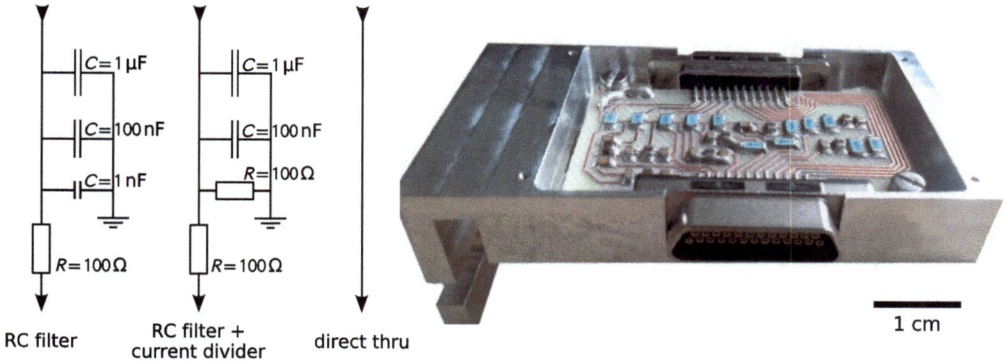

Figure 4.3: Filter box for DC current bias. Schematics (left) and photograph (right) of the DC bias filter box at the 3.5 K stage. The filter is divided into 3×8 wires, where the ones on the left only have a *RC* filter, the ones in the center have an *RC* filter and a 1 : 1 current divider and the ones on the right simply pass through. Note that the image only shows the upper side of the layout and the bottom side looks the same. For the *RC* filters, multiple capacitors are used in parallel, as the ones with a high capacitance have a bad high frequency transmission.

4.1.4 DC bias

To be able to tune the transition frequency of the tunable sample described in Sec. 4.3, a current has to be provided down to millikelvin temperatures. The focus here lies not on the high current regime but on additional filtering, which removes high frequency noise and provides a constant current as stable as possible. The wiring between room temperature and the 3.5 K stage is made from BeCu, the rest down to the base stage by NbTi. All wires are multi filament braided looms, where adjacent lines are twisted to reduce coupling of electromagnetic noise. To filter out noise, a filter box was designed for the 3.5 K stage, shown in Fig. 4.3. This box features three different filter configurations, to allow for a flexible usage. While one part is only equipped with *RC* filters, the second features an additional current divider to divide the current and with that the current noise on low temperatures.

4.2 Microwave measurements

4.2.1 Spectroscopic measurements

For many characterization experiments of superconducting resonators and qubits, it is sufficient to probe the system in the steady state. Using commercially available devices which are designed for this task, a standard spectroscopic setup can be assembled from few parts. Coming along with this structural simplicity, also the control and data acquisition is very simple with this setup.

A typically used setup for spectroscopic measurements is shown in Fig. 4.4. It consists of a vector network analyzer (VNA) as readout device which probes the resonance spectrum of the resonator and additional microwave sources to apply probe and drive tones. To reach a high drive amplitude despite the high attenuation on the input lines, we use a high power microwave amplifier at room temperature for the measurements presented in Ch. 5. The choice of attenuation at the output of VNA and microwave source shows a typical configuration and is adapted to the needs of the experiment, respectively.

4.2.2 Time resolved measurements

We use a time domain setup to measure the characteristic timescales of the qubit, which is described in the next section. Beforehand, we will shortly get an overview, which physical quantities can be measured by which pulse sequence.

In general, all time domain measurements consist of a sequence of pulses, after which a readout tone is applied. The readout projects the qubit state onto the z axis in the Bloch sphere and therefore on one of the states $|0\rangle$ or $|1\rangle$. This pulse sequence is then repeated for a number of different pulse lengths or delay times, depending on the protocol. To acquire statistics of the measurement, this protocol is then repeated several thousand times and the result is recorded. It is important to repeatedly loop through all pulse configurations instead of directly repeating each configuration multiple times, to ensure that fluctuations and slow drifts affect all acquired data points in the same way. After the readout, a waiting time before the next pulse is implemented to let the qubit and resonator relax back to their ground states. The three-dimensional representation of the complex qubit state was performed by quantum state tomography [77].

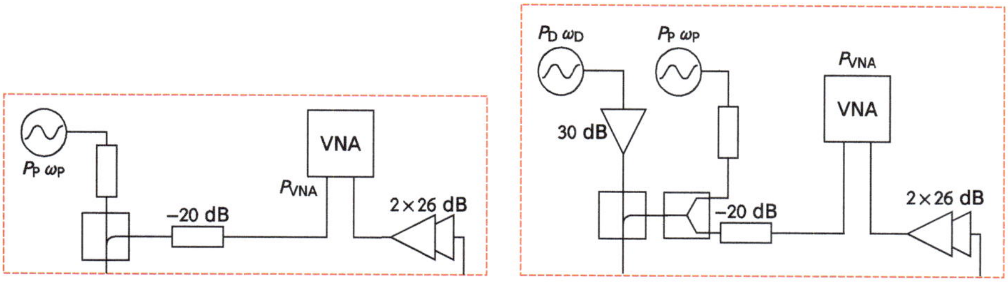

Figure 4.4: Spectroscopic setups. A two-tone (left) and three-tone (right) microwave setup for spectroscopy measurements. Both setups include a vector network analyzer (VNA) to probe the resonator and a probe microwave source to probe the qudit level transitions at different powers and frequencies, P_P, ω_P. In the three-tone setup, an additional drive microwave source with an high power amplifier is included. The attenuation indicated in the scheme represents typical values and is adjusted depending on the experiment.

Rabi measurement

The Rabi measurement consists of an excitation pulse at the qubit frequency, with increasing pulse length τ, followed by the readout. We recall from Eq. (2.8) that this corresponds to applying $\hat{\sigma}_x$ for the time τ and the qubit state starts to rotate around the x axis. This can be seen in Fig. 4.5 a) as a cosine-shaped oscillation in the readout value. Rabi measurements are frequently performed to measure the time t_π, for which the pulse of distinct power has to be applied to rotate the qubit state from $|0\rangle$ to $|1\rangle$. The readout signal is then expected to be

$$\langle \hat{\sigma}_z \rangle = e^{-\tau/T_{\mathrm{Rabi}}} \cos(2\pi\tau/2t_\pi), \tag{4.1}$$

with $T_{\mathrm{Rabi}} = 2/(T_1^{-1} + T_2^{-1})$.

Relaxation time measurements

To measure the relaxation time T_1, a π pulse is applied to bring the qubit in the $|1\rangle$ state. After waiting for a time Δt, the remaining qubit excitation is read out. Since the qubit does not acquire any phase information in the relaxation process, the trajectory on the Bloch sphere goes directly through the origin of the sphere, see Fig. 4.5 b). If the initial π pulse did not perfectly match in length or frequency, the initial state is different, but we still observe an exponential decay of

$$\langle \hat{\sigma}_z \rangle = 1 - 2e^{-\Delta t/T_1}. \tag{4.2}$$

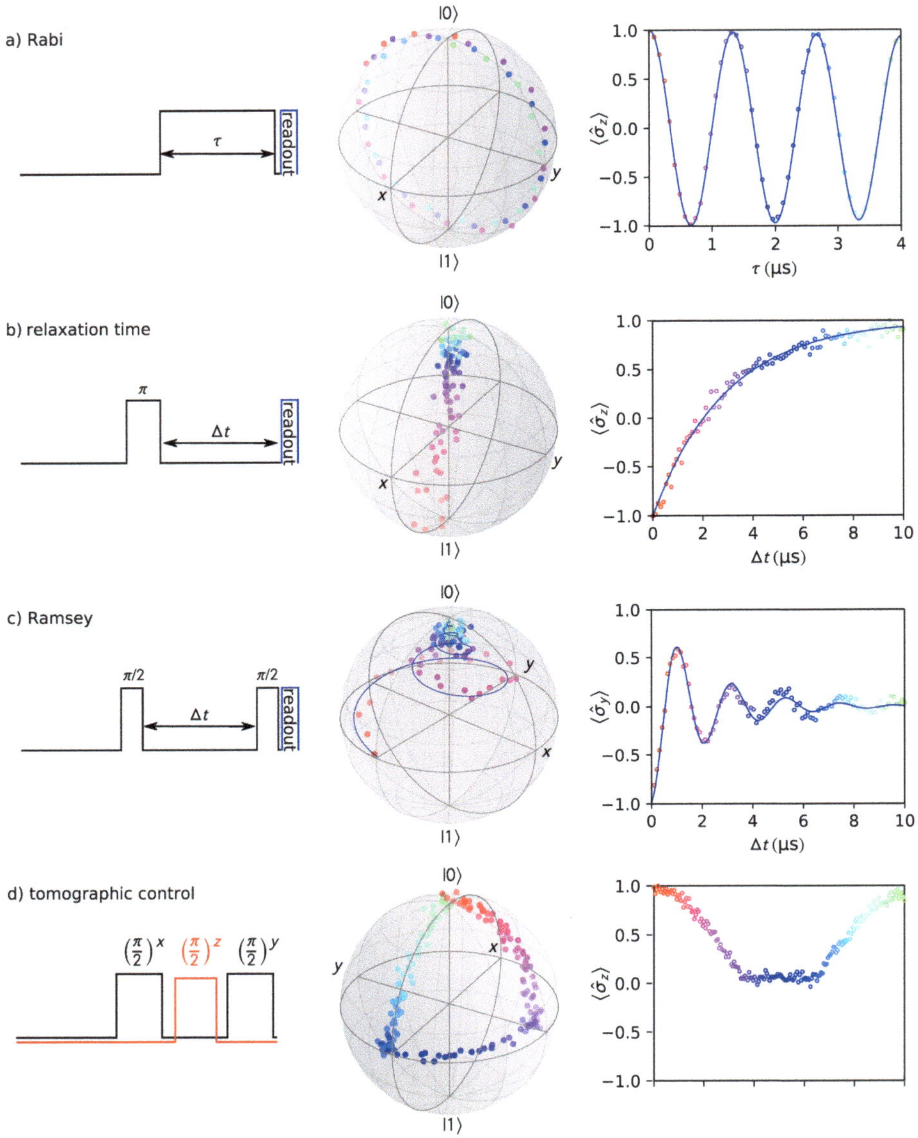

Figure 4.5: Typical time domain measurements. Each subfigure shows the pulse sequence at first, a tomographic image of the qubit state evolution in the center and the typical measurement outcome on the right. The time in the data points is color coded, from red to turquoise. a) A Rabi experiment which rotates the qubit state around the x axis. b) For the relaxation time measurement, the qubit is prepared in the $|1\rangle$ state and the time evolution is recorded. It is well visible that no additional phase information is acquired by the decay. c) A Ramsey experiment with a detuning between drive and qubit frequency. The Bloch sphere in the center shows the evolution before the second $\pi/2$ pulse is applied, which enables to measure the value of $\langle\hat{\sigma}_y\rangle$. d) Demonstration of an all-axis control of a fast tunable sample. All measurement data and plots are taken from [77].

Ramsey measurement

For a Ramsey measurement, the qubit is prepared in the state $(|0\rangle + |1\rangle)/\sqrt{2}$ by applying a $t_\pi/2$ pulse. After a time Δt, a second $t_\pi/2$ pulse is applied and the state is read out. If the qubit state is unperturbed in the time Δt, we will always measure the $|1\rangle$ state. From Eq. (2.8) we know however that even if no drive is applied, the operator $\Delta_\omega \hat{\sigma}_z$ effectively acts on the qubit state, where Δ_ω is the detuning between drive and qubit frequency. Also, any relaxation or dephasing will affect our measurement outcome, resulting in $|0\rangle$ before the last $t_\pi/2$ pulse. This measurement scheme is depicted in Fig. 4.5 c) and follows the following formula:

$$\langle \hat{\sigma}_z \rangle = -\mathrm{e}^{-\Delta t/T_2^{\mathrm{R}}} \cos \Delta_\omega \Delta t, \tag{4.3}$$

where we explicitly keep the index R on the characteristic timescale T_2^{R} to avoid any ambiguity and confusion of the star notation.

To measure the Ramsey dephasing time, it is typically beneficial to not set $\Delta_\omega = 0$, but chose a small detuning such that several oscillations are visible. This gives a much clearer envelope and the possibility to distinguish between frequency mismatch and dephasing. The detuning Δ_ω is technically not the actual detuning between drive and qubit frequency, but the rate $\Delta\varphi/\Delta t$ at which the phase between the two $t_\pi/2$ pulses is adjusted.

Due to the oscillating nature of this measurement result, Ramsey measurements are commonly used when a sensitive measurement of any change in qubit frequency is needed. If the T_2^{R} time is long enough, the time within which the qubit can accumulate phase and therefore the resolution of the Δ_ω measurement grows.

Hahn echo measurements

If we assume that the dephasing is caused by slow changes in the qubit transition frequency, this means that Δ_ω is different for each measurement and the averaged vector on the Bloch sphere has zero length after short Δt when averaging over some seconds. This can be compared to averaging the outcome of multiple Ramsey experiments with long T_2^{R} times but different frequency, which will result in a very fast decay of the resulting envelope.

If we assume that the fluctuations are so slow that Δ_ω does not change between the first pulse and readout, we can counteract these fluctuations by applying a t_π pulse in the middle of the waiting time Δt. This pulse flips the qubit state from $(|0\rangle + |1\rangle)/\sqrt{2}$ to $(|0\rangle - |1\rangle)/\sqrt{2}$, which effectively changes the sign of the acquired phase. The phase acquired after the t_π pulse then compensates this and independent from a Δ_ω, the

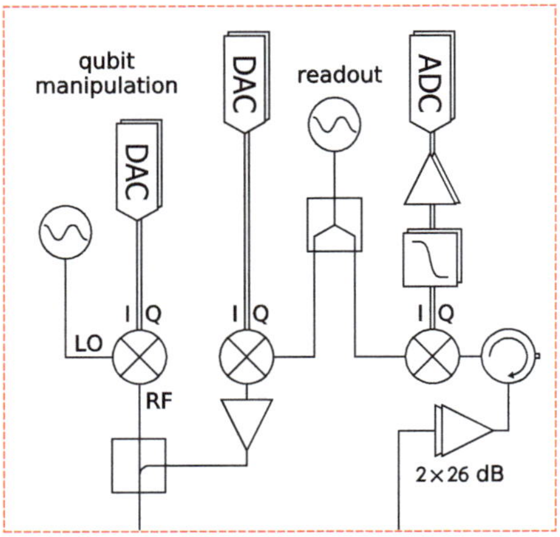

Figure 4.6: The time domain setup. To enable pulsed measurements, a four channel digital-to-analog converter (DAC) is used, where each pair of channels creates the pulse envelopes for the IQ mixers. By using the same microwave source for up and down conversion of the readout signal, we ensure a common frequency and phase reference. The signal is recorded by a two-channel analog-to-digital converter (ADC).

state will end up in $(|0\rangle - |1\rangle)/\sqrt{2}$. Faster fluctuations in Δ_ω cannot be compensated by this type of experiment. Due to the independence from the absolute value of Δ_ω, the measurement signal shows no oscillation but a decay following

$$\langle \hat{\sigma}_z \rangle = \mathrm{e}^{-\Delta t / T_2^{\mathrm{E}}}. \tag{4.4}$$

The name "echo" measurement comes from NMR experiments, where a spin ensemble is used. In this case, all spins oscillate in phase again after $\Delta t/2$ after the t_π pulse, such that they emit a measurable microwave signal. This signal can be seen as echo of the initial excitation pulse.

4.2.3 Time domain setup

To perform the measurements introduced in the last section and therefore measure the characteristic timescales of the qubit, we use a time domain setup, which is depicted in Fig. 4.6. The pulse envelopes are generated by a four-channel arbitrary waveform generator (AWG) and fed to the I and Q inputs of two IQ mixers, one for the qubit manipulation pulses and one for the readout pulse on the resonator frequency. The two signals are then combined by a directional coupler. The RF inputs for the mixers are generated by two local oscillators (LO). Since IQ mixers are nonlinear and very non-ideal components, the technique of single sideband mixing is utilized, where the pulse envelopes are multiplied by a sine and cosine signal and fed to the I and Q inputs of the mixer, respectively. The oscillating signals typically have a frequency of $\omega_{IQ}/2\pi = 63\,\text{MHz}$ for the readout channel and $\omega_{IQ}/2\pi = 75\,\text{MHz}$ for the manipulation channel. This shifts the generated tones in frequency, which means that we can shift the frequency of the LO by the IQ frequencies and any leakage of the LO due to non-ideal mixers will not be resonant with any transition on the qubit chip. Constraints for these IQ frequencies are the output bandwidth of the AWG, the bandwidth of the IQ mixers and the anharmonicity of the qubit, since we do not want to excite any other transition with a leaking LO. Also, the IQ frequencies should be different from each other to account for possible cross-talk of the AWG channels. The generation and optimization of phase-controlled pulses with this setup was in-depth described in Ref. [77].

The reflected readout signal from the resonator is amplified and down-converted by an IQ mixer with the same reference LO as for the up-conversion. After a low-pass filter to eliminate high-frequency mixing products, the signal is amplified and fed into a two-channel digitizer. The acquired data for multiple iterations of the pulse sequence are averaged on a computer and the amplitude and phase value of the complex fast Fourier transformation at the IQ frequency of the readout channel is calculated. This result corresponds to the S_{11} signal of the resonator at a fixed frequency and can be used to detect a shift of the resonator resonance frequency and therefore the qubit level population.

4.2.4 Data evaluation and measurement software

For the experiments throughout this thesis, a lot of effort has been spent to enable and facilitate the fast and easy acquisition of measurement data, alongside with a configurable data viewer, fast evaluation and fitting tools and a flexible database to organize all measurement data. This software package, named QKIT, is written in python and freely available under Ref. [141].

This toolset enables to repeatedly acquire spectroscopic or time domain data, where almost any imaginable automation can be included. Allowing for an automated fit after the measurement, the calibration data of a Rabi experiment can for instance be automatically fed to the next T_1 measurement, which requires the t_π time for an excitation pulse. Especially for the magnetic measurements, it was from time to time necessary to thermally cycle the sample, i.e., bring it over the critical temperature for a short time to eliminate all trapped flux. This procedure can be easily included into the measurement scheme and the measurement automatically continues, when a satisfactory base temperature is reached again. Finally, the measurement sequence automatically logs and saves all settings of the measurement devices, leading to a semi-automated lab book and reproducible measurement schemes.

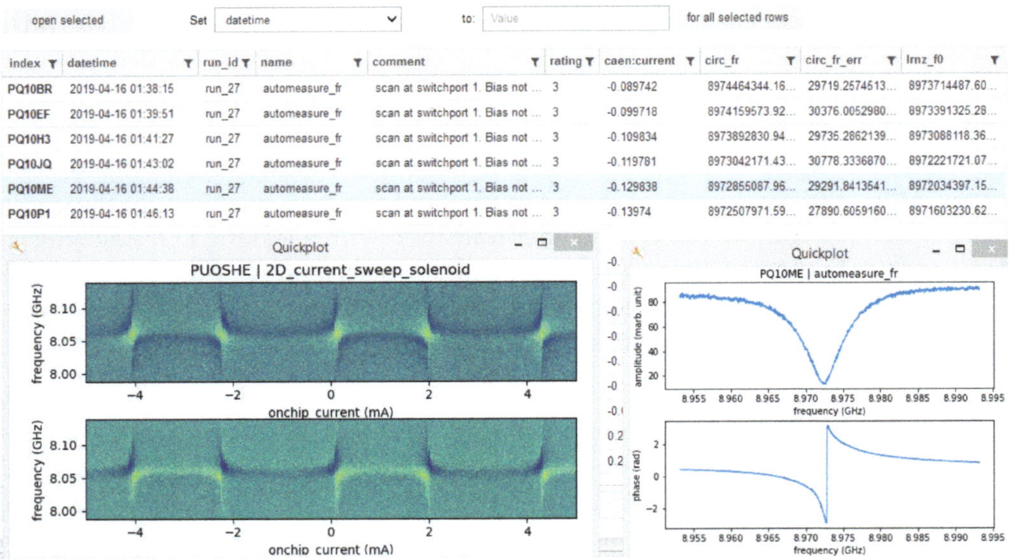

Figure 4.7: The measurement database and `quickplot` feature of qKIT. Above, an excerpt of the measurement database is shown which lists after the unique ID of each measurement the starting timestamp, comments, a user defined rating, relevant device parameters and fit results of the data. All user-defined data can be directly changed in the table and sorting and filtering can be enabled to restrict the range of listed measurements. On the bottom, two windows of different types of measurements are shown. To the left, the anti-crossings of the tunable transmon with the resonator frequency are shown in amplitude and phase as color coded data. Any constant background is eliminated in this view by subtracting the average over the x direction. While this coarse method is not directly applicable for pictures intended for publications, it increases the contrast of the measured data and allows to clearly see even small changes. To the right, a reflection measurement of a resonator is shown.

When acquiring lots of data by varying different tuning parameters, it is extremely helpful to be able to quickly screen and review the recorded data to see relevant features and be able to scan interesting regions in more detail. A recently developed `quickplot` function enables to plot the relevant data for each measurement in the database within fractions of seconds (depending on the size of the acquired data) and automatically apply filters for background removal, phase wrapping or elimination of runaway values. Each measurement can be identified with a six-digit unique ID and the meta data of all measurements is stored in a measurement database. This database allows to filter for samples, measurement type, fitted values, or any other parameter and enables to give ratings for the quality and relevance of a measurement. In sum, these tools help to keep an overview over all acquired data and find relevant measurements again quickly. A demonstration of this database and the `quickplot` interface can be seen in Fig. 4.7.

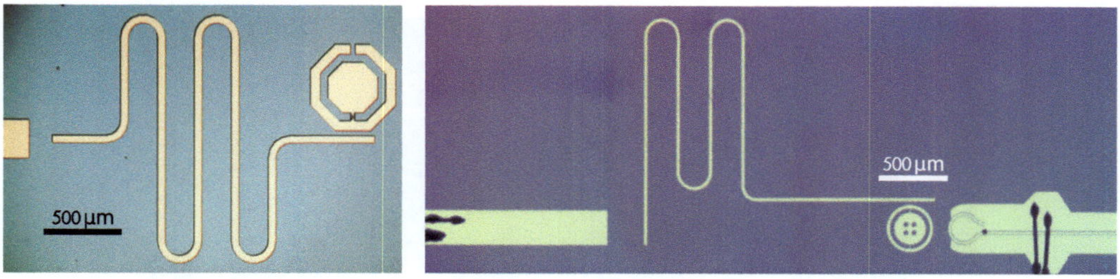

Figure 4.8: The two qubit samples used throughout this thesis. Micrographs of the two types of transmon qubits. On both images, from left to right, the microstrip microwave transmission line, the readout resonator and the qubit is shown. Left: The non-tunable transmon made from TiN, fabricated at NIST. Right: The gradiometric concentric transmon is in principal tunable by a current flowing through the flux bias loop on the right. This loop induces unequal fluxes in the qubit's tuning loops and therefore tunes the qubit, as described in Sec. 2.5.1. The flux bias loop is executed as coplanar waveguide structure with a ground plane directly next to the conductor to reach a 50 Ω microwave impedance. To ensure a proper grounding, two bonds are connecting both sides of the ground plane.

4.3 Qubit samples

For the experiments in this thesis, two different versions of the concentric transmon have been used, which also differ in choice of material. A micrograph of the two samples is given in Fig. 4.8.

The first chip was fabricated at the NIST[3] by the group of Martin Sandberg and David Pappas, and gratefully delivered to us for measurements. Besides the suppression of the dipole moment as described in Sec. 2.5.1, the focus for this sample was the use of titanium nitride, a material engineered for a high critical temperature [142] and low microwave losses [143]. TiN has the additional advantage to be stable in humid air at room temperature and is therefore not likely to form surface oxides. The Josephson junction is however realized by a $Al/AlO_x/Al$ structure, which has been found to give the best coherence. This two-step process brings additional features which are explained in Ch. 6.2.2.

From spectroscopic measurements, we find the fundamental qubit transition frequency as $\omega_{01}/2\pi = 4.755\,\text{GHz}$ and determine from the higher level transitions a charging energy of $E_C/h = 197.7\,\text{MHz}$ and a Josephson energy of $E_J/h = 15.5\,\text{GHz}$. This results in a ratio of $E_J/E_C = 78.4$. The full spectrum of this qubit can be found in Fig. 4.9.

[3] National Institute for Standards and Technology in Boulder, Colorado

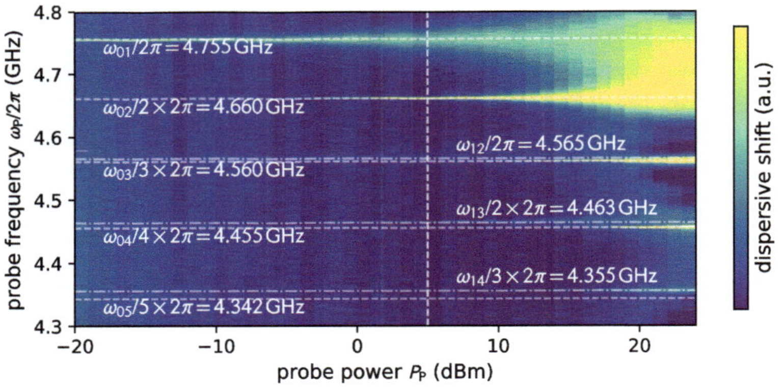

Figure 4.9: Transition frequencies of the TiN transmon sample. Results from a two-tone experiment to see the qudit level transitions under increasing qudit probe power. Since multi-photon transitions start to occur at higher powers, the transitions can be identified and E_J and E_C can be extracted. Due to an increased sample temperature, also transitions starting from the $|1\rangle$ state can be seen (dash-dotted lines), which fit nicely to the values expected from the qubit parameters. Adapted from [71].

Capacitively coupled to the qubit is a $\lambda/2$ resonator at $\omega_R = 8.573\,\mathrm{GHz}$ by a coupling strength of $g/2\pi = 71.5\,\mathrm{MHz}$. On its other end, the resonator couples to a $50\,\Omega$ microwave transmission line in microstip design.

The second type of sample was fabricated at KIT by Jochen Braumüller and is in-depth described in Ref. [99]. As explained in Sec. 2.5.1, this design features two JJs in a gradiometric SQUID geometry, which makes the qubit transition frequency sensitive to field gradients but not to homogeneous fields. For the experiments in Ch. 7 at higher magnetic field values however, we found that the inhomogeneity of the magnetic field is large enough to significantly tune the qubit frequency. Together with a fluctuating current of the solenoid's power supply, the qubit transition was strongly broadened and slowly drifted with time. This made the qubit unusable for magnetic field sweeps with the given magnetic setup. The experiment was therefore executed with a non-tunable transmon sample, where one of the JJs was open due to fabrication issues. This qubit sample had a fixed frequency, which only weakly changes in the magnetic field. Since the exact same qubit design was used as for the tunable qubit, all experiments in Ch. 7 can be easily repeated with a tunable transmon once the bias magnetic field is stabilized.

The transition frequencies of this sample are shown in Fig. 4.10, from which we can calculate a charging energy of $E_C/h = 182\,\mathrm{MHz}$ and a Josephson energy of $E_J/h = 32.7\,\mathrm{GHz}$. For this sample, we get $E_J/E_C = 179.7$, being even deeper in the transmon regime than the TiN sample. In addition to a $\lambda/2$ readout resonator at $\omega_R = 8.78\,\mathrm{GHz}$ which is also coupled to a microstrip transmission line, this

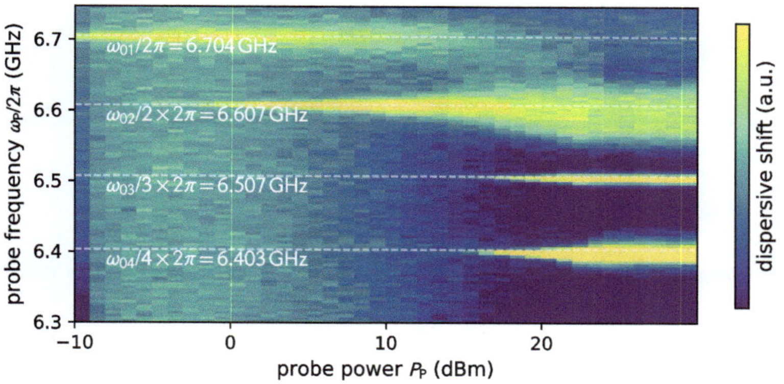

Figure 4.10: Transition frequencies of the all-Al qubit sample. Results from a two-tone experiment to see the qudit level transitions under increasing qudit probe power. We again see the multi-photon transitions starting at higher probe powers. Due to the high $E_J E_C$ values extracted from this data, this sample is even deeper in the transmon regime than the TiN sample, which can be seen in an almost constant frequency difference between adjacent multi-photon transitions.

design includes a 50 Ω impedance matched flux bias line to be in principle able to tune the qubit by fast pulses.

4.4 Sample fabrication

4.4.1 Qubit samples

Since for this work, a variety of samples generously provided by other people was available, there was no need to fabricate own qubit chips. To understand the interference effects presented in Ch. 6 better, the fabrication process is sketched shortly here. The fabrication procedure of the JJs with shadow angle evaporation is shown in Fig. 4.11. For the TiN chip, a two-stage process is used, where first the resonator, transmission line and qubit islands are structured in TiN by optical lithography. In the Al chip, all structures including the junctions are integrated in one layout. The resist system for electron beam lithography consist of two different resists, which have different sensitivity to electrons. It is therefore possible to create an undercut under a free standing bridge. In a first step, aluminum is evaporated under an angle, where no material is deposited in the shadow below the bridge. In the second step, a controllable partial pressure of oxygen is applied to the chamber, which oxidizes the surface of the aluminum. Next, Al is evaporated under a different angle, which leads to a small overlap contact of the two Al layers. Finally, the resists are removed, which takes away all unwanted metalization.

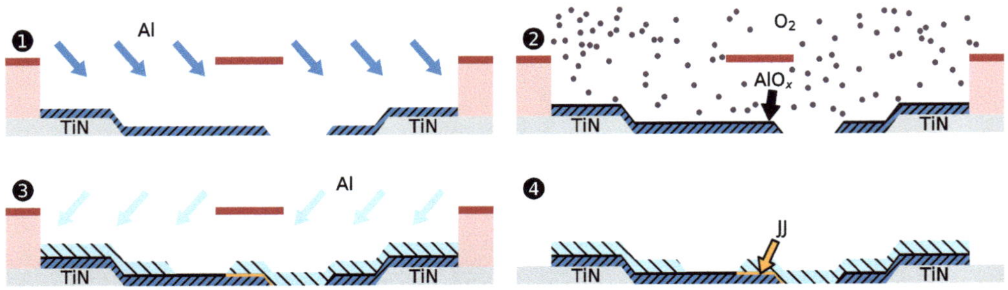

Figure 4.11: Fabrication process of Josephson junctions. In a preceding step, the large metal parts are created from TiN and the two resists (light and dark red) are structured. The resists form a free standing bridge which is being hold in the third dimension not shown here. First, aluminum is evaporated under an angle (blue arrows), whose surface is oxidized in the second step by an oxygen atmosphere. Afterwards, a second layer of Al is deposited (blue arrows), where a small overlap contact between the two layers is formed (yellow area). After removing the resists, the device is ready to be used and can be contacted. The other Josephson contacts (black areas in step 4) are commonly neglected due to their large size but become relevant in Ch. 6.

4.4.2 Magnetic samples

The magnetic material used in Ch. 7 was grown by sputter deposition with synchronous co-sputtering of Iron and Cobalt. To get a stoichiometry of about $Fe_{0.7}Co_{0.3}$, the sputter guns are adjusted to have a power of $P_{\mathrm{Fe}} = 100\,\mathrm{W}$ and $P_{\mathrm{Co}} = 37\,\mathrm{W}$, given by measured individual deposition rates and known material densities. After this process, a protective coating of AlO_x was sputtered to prevent a fast oxidation of the cobalt-iron alloy in the further processing steps and measurements. The fabrication parameters for the optical lift-off process are given in Appendix B.

After structuring the ferromagnet, a second optical process was applied to structure rectangles of resist at the edges of the individual chips. These structures are baked on a hot plate at 230 °C for 6 minutes to glass the resist, making it insoluble in solvents. Additional parameters to this process are also given in Appendix B.

5 Sensing microwave fields with superconducting qubits

Sensing a microwave signal directly on the quantum chip at millikelvin temperatures is a challenging task, especially when the photon energy is widely undefined. In this chapter, I present a method to perform these measurements based on the AC Stark effect on a multi-level superconducting quantum circuit. After a general introduction to quantum sensing, I present the AC Stark effect first in general and then in detail for a multi-level system by means of analytic and numeric theory. After a comparison with measured data, I detail on the sensing scheme and present a first proof-of-principle experiment, leading to more advanced measurements. The results of the experiments in this chapter lead to the publication of Ref. [71].

5.1 Quantum sensing

Although quantum technologies are used for sensing and measurement devices already for several decades, the broad field of quantum sensing is still highly subject to active research and development. An outstanding example in the field of superconductivity is the concept of a SQUID (see Sec. 2.4.1), which was developed over 55 years ago [144] and is still very frequently used in various fields of research, manifested by its appearance in more than one publication per week[1] on average. Similar to semiconducting sensors which had applications long before semiconducting computers became relevant, it becomes more and more apparent that quantum sensors are of general use already before quantum computers are considered to be state of the art. The amenities of the development of new quantum sensors compared to quantum computers is that all spurious and unwanted effects in quantum computing, leading to noise, decoherence, and gate errors, can in principle be used to build a quantum sensor which exactly measures these decoherence effects and spurious influences. The weakness of quantum computing can therefore be seen as the strength

[1] Source: Preprints published on arXiv.org, based on the data for Jan-Nov 2019.

of quantum sensing. This means, on the other hand, that the emerging topic of quantum sensing can be and frequently is used to reuse spin-off results from research on quantum computers, where a restriction in the applicability of quantum computing was found and turned into a feature for quantum sensing. Being a valid approach for the development of quantum sensors, this motion in the field blurs out the name "quantum sensing". Although being not clearly defined, usually this term is used when one of the following three features is used to measure a physical quantity [29]:

1. A quantum object,
2. quantum coherence or
3. quantum entanglement.

While strictly speaking only the third point really exploits the quantum nature of the sensor and opens the possibility to reach a sensing resolution beyond the classical limit, the other types of sensors also have their right to exist and are commonly entitled as quantum sensors. Despite not being "super-sensors", they are still able to beat the resolution and flexibility of classical and established sensors. Especially when examining other quantum systems, they can eventually be the only sensors applicable in terms of size, energy scale and coupling mechanisms.

The list of quantum sensors which are nowadays routinely used is very long, and some of them are not even recognized as quantum sensors any more, since they are so commonly used. One very basic example is the usage of reference substances for ESR and NMR[2] experiments, which have a characteristic frequency-to-magnetic-field relation and sharp resonance features. This makes them useful as magnetic field reference sensors, frequently used to calibrate the magnetic field created by permanent magnets or a magnet coil.

Speaking of NMR experiments, the whole human body can be turned into a quantum sensor in the context of magnetic resonance imaging (MRI). Here, the NMR properties of the hydrogen atoms in the body are measured and spatially resolved, giving insight to the chemical environment of the H atom. Measuring T_1 and T_2 makes it possible to draw conclusions on the type of tissue, water and fat contents, and more. By using special techniques it is also possible to characterize motion and flow as well as metabolic processes in the body in real time, making MRI to a valuable examination tool in medicine [145].

One of the most broadly used types of quantum sensors with respect to sensing quantities and application areas are nitrogen vacancy (NV) centers, which represent

[2] ESR: electron spin resonance, NMR: nuclear magnetic resonance

electronic spin defects in the crystal structure of diamond. They have been demonstrated to be able to measure magnetic fields, electric fields, pressure, and temperature and can be used as bulk material or in the form of stable single nano crystals. As nano crystals, they can be added to a scanning probe, which allows for sub-100 nm resolution in magnetic field resolution of surfaces. Or they can even be inserted into living cells where they allow for an *in-vivo* temperature measurement [29].

Similar to that, SQUIDs can be mounted onto a scanning tip where they are used for measurements of magnetic field, current and temperature with ultra high spatial resolution. By fabricating the nanoSQUID directly onto the tip, SQUID loop diameters of less than 50 nm have been demonstrated [146, 147].

Exploiting not only quantum effects but also the sensitivity and vulnerability of a coherent quantum state, superconducting quantum bits can also be used as quantum sensors and tools for investigation. Compared to the previously presented nuclear or electron spins, superconducting qubits are macroscopic quantum objects that can have large magnetic and electric dipole moments and are therefore particularly suitable for coupling and sensing experiments. In a great variety of experiments, they have been used to study for example environmental noise input or spurious coupling to defects by means of T_1 relaxometry or resonant coupling [30, 36, 148].

To study an oscillating physical system of interest by means of a superconducting qubit, usually strong coupling is desired, which enables a coherent exchange of excitation and therefore information about the system. The necessary coupling strengths can however be challenging to achieve and some effects are only visible when the two systems are close in resonance, like the level repulsion described in Sec. 3.6.

Instead of a coherent excitation exchange, it would be beneficial to be able to use the qubit to sense the oscillation of the second system, which would reduce the severity of the second system's linewidth and therefore the need to be in the strong coupling limit.

Being able to detect arbitrary microwave signals by a qubit would also open new ways to characterize and optimize environments for quantum experiments, especially with respect to quantum computation. This sensing scheme would allow to directly measure the cross-talk from all other elements on a quantum chip to the qubit, which could otherwise only be characterized in the resonant case or by means of simulations. Likewise, it is practically impossible to measure the full frequency dependent transmission from the microwave source to the qubit position by other means, as the wiring in the cryostat as well as the transmission on the quantum chip change when cooling down from room temperature to several millikelvin.

To construct such a microwave sensor from a superconducting qubit, we will in the following study the influence of an oscillating field on a quantum bit. This is generally described by the AC Stark effect, which will be explained and detailed in the next sections.

5.2 The AC Stark shift

In analogy to the Zeeman effect which splits atomic transitions in a magnetic field, Johannes Stark studied the influence of a static electric field on these transitions [149] and Schrödinger later derived a theoretical treatment for this effect [150]. It is since understood as the splitting and shift of energy levels from the interaction of a molecule's dipole moment with the electric field. To observe this effect, strong electric field strengths are necessary, being experimentally hard to achieve and control. The DC Stark effect was therefore not as extensively studied as its magnetic counterpart, the Zeeman effect.

With the introduction of lasers however, it was possible to strongly increase the limits of experimentally achievable AC electric field strengths and hence study the AC Stark effect. Experimentally, the shift and splitting of the D absorption line in potassium under a strong laser drive could be observed [151], and many other experiments followed including atoms or molecules in an optical cavity.

With the formation of circuit quantum electrodynamics (QED) in analogy to cavity QED, it was obvious that also an artificial atom, i.e., a superconducting qubit, should show a shifting of its energy levels when strongly populating the coupled readout resonator [76], which was then also experimentally observed [152].

In both atomic QED and circuit QED however, the term "AC Stark effect" is sometimes generalized to all experiments including a strong drive tone or field [153], resulting in the description of Rabi or Autler-Townes splittings [97]. In this thesis, we will focus instead on non-resonant drives and study their influence on the qubit transitions.

5.2.1 Resonator-induced AC Stark shift

Following the analogy to cavity QED, the AC Stark shift is commonly studied as the influence of a coupled resonator. In many publications, it is therefore seen as the coun-

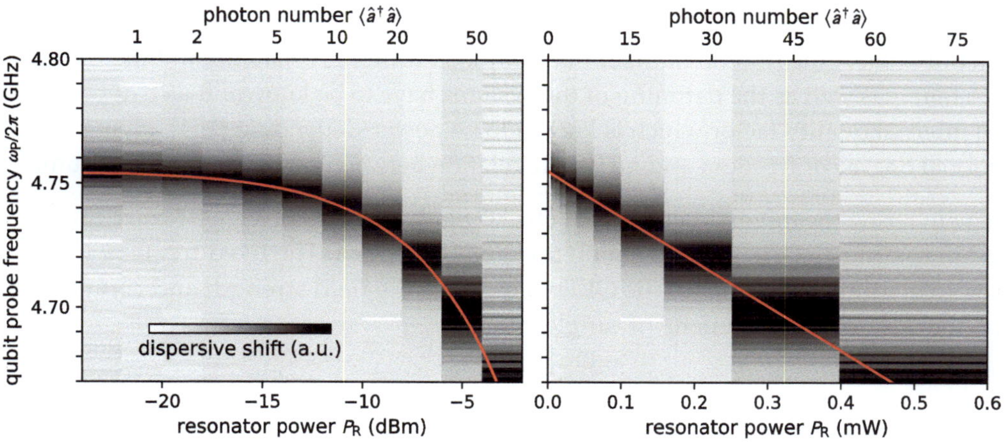

Figure 5.1: AC Stark shift caused by resonator population. We populate the readout resonator with different photon numbers by increasing the power P_R at the VNA to read out the resonator. The red line is a fit of the qubit transition frequency to the data after Eq. (5.1) from which we expect a linear power dependence. This can be seen in the right plot, where the power is given in linear power units (mW). The left plot with the power given in logarithmic units (dBm) however renders the low photon number regime in greater detail and is therefore typically used. The photon number indicated in the upper axes is found by $\Delta\omega_q = \langle\hat{a}^\dagger\hat{a}\rangle\, g^2/\Delta_r$, where $g/2\pi = 71.5\,\text{MHz}$ and $\Delta_r/2\pi = 3.818\,\text{GHz}$ is known from other experiments. This means a single photon shifts the qubit frequency by $\Delta\omega_q = 1.3\,\text{MHz}$. We also see that the single photon regime can be reached by setting the output power of the VNA to $P_r \approx -21\,\text{dBm} = 7.5\,\mu\text{W}$. Adapted from [71]

terpart to the Lamb shift, presented in Eq. (2.11) for the qubit readout [76]. By simply shifting the assignment of the $\chi\hat{\sigma}_z\hat{a}^\dagger\hat{a}$ term to the qubit part, the effect is directly obvious:

$$\hat{H}_{\text{eff}}/\hbar \approx (\omega_r + \chi\hat{\sigma}_z)\,\hat{a}^\dagger\hat{a} + \frac{1}{2}(\omega_{01} + \chi \qquad\qquad)\,\hat{\sigma}_z \tag{2.11*}$$

$$\approx (\omega_r \qquad\quad)\,\hat{a}^\dagger\hat{a} + \frac{1}{2}(\omega_{01} + \chi + \chi\hat{a}^\dagger\hat{a})\,\hat{\sigma}_z. \tag{5.1}$$

Now, a shift of the qubit frequency $\Delta\omega_q = \chi\hat{a}^\dagger\hat{a}$ proportional to the number of photons in the resonator can be seen, where we recall that the dispersive shift is given by $\chi = g^2/\Delta_r$.

This behavior is experimentally demonstrated in Fig. 5.1, where the power of the readout signal was steadily increased, resulting in an increased number of photons in the resonator and eventually in a shift of the qubit transition frequency. We recall from Eq. (3.6) that the population in the resonator is given by $\langle n\rangle = P_{\text{in}}\frac{4}{\hbar\omega_0^2}\frac{Q_L^2}{Q_c}$. This dependence is commonly used to adjust the power level for the readout tone to be in the single photon regime and to reassure that the qubit transition frequency is not influenced by the readout.

For real sensing applications however, the effect in this form is not of great use, as we can only detect the photon number of a coupled harmonic oscillator and the coupling constant g as well as the detuning of the systems have to be known. It also depends on the internal quality factor which is known to be power-dependent [154], giving an additional degree of freedom and no usable relation between power and photon number.

Remembering the initial description of the AC Stark effect however, we realize that only an alternating electric field acting on the atom is needed. The fixed frequency cavity was solely introduced in atomic QED to increase the field strength and compensate for the low dipole moment of single atoms.

5.3 Multi-mode AC Stark shift

To access a more generalized approach to the AC Stark shift than the cavity-atom-model, we will expand our analysis in the rest of this chapter by the following factors: Instead of a resonant cavity system, we will treat a more universal drive field, which experimentally does not limit us to pre-defined elements or frequencies and lets us describe a greater variety of systems and processes. Also, the atomic part of the system will no longer be restricted to a true two level system, but we allow a more universal multi-level system with a small anharmonicity. This gives a better description of the system we are using experimentally and finally makes more observables accessible, which can later be used for an enhanced sensing scheme.

In the following, we will access this regime using two different theoretical approaches: First, we start with a theoretical analysis of the system. Using a simplified model for our anharmonic system and applying the approximations of perturbation theory, we find an analytic formula for the qudit transition frequencies under the influence of the drive. This relation provides additional insights and allows us to derive simplified expressions for special cases. Due to the implied approximations however, these solutions cannot fully describe our experimental observations.

For this reason we utilize numerical simulations in a second step, including a more detailed model of our qudit system and excluding the approximations of perturbation theory. Yielding good results on the exact problem we want to solve, this method however comes with the drawback of only calculating a solution for the given starting parameters, without any gain in physical understanding. Both paths therefore have their own benefits and will be presented in the following.

5.3.1 Approximate analytical solution

To derive an analytical expression for the qudit transition frequencies under the influence of a drive, we start with the approximated transmon Hamiltonian of Eq. (2.29), where a constant anharmonicity is assumed:

$$H_{\mathrm{q}}/\hbar = \omega_{\mathrm{q}}\hat{b}^{\dagger}\hat{b} - \frac{\gamma}{4}\hat{b}^{\dagger}\hat{b}\left(\hat{b}^{\dagger}\hat{b}+1\right). \tag{5.2}$$

For simplicity, we introduced the harmonic fundamental transition frequency $\omega_{\mathrm{q}} = \sqrt{8E_{\mathrm{C}}E_{\mathrm{J}}}/\hbar$ and the anharmonicity $\gamma = 2E_{\mathrm{C}}/\hbar$ and neglected the constant terms. We model our drive field with amplitude A_{D} and frequency ω_{D} by the Hamiltonian

$$H_{\mathrm{D}}/\hbar = A_{\mathrm{D}}\left(\hat{b}+\hat{b}^{\dagger}\right)\cos\omega_{\mathrm{D}}t.$$

The amplitude of the drive A_{D} is here given in units of an angular frequency, which corresponds to the usage of the Rabi drive amplitude A in Eq. (2.9). On resonance, this is the frequency of the observed Rabi oscillations, meaning that the qubit changes from $|0\rangle$ to $|1\rangle$ within the time $\Delta t = \pi/A$ and the energy of one $\hbar\omega$ is then stored in the qubit. To satisfy an experimental physicist's understanding, a simple power-to-amplitude relation can hence be given by $P = A\hbar\omega/\pi$.

The beauty of choosing the amplitude for our analysis is that it describes the power directly applied to the qudit, independent of any coupling or frequency-dependent transmission function. As we will later use the readout resonator only to probe the qudit transition frequencies, we can neglect it here for our analysis. The full Hamiltonian then reads:

$$H/\hbar = \omega_{\mathrm{q}}\hat{b}^{\dagger}\hat{b} - \frac{\gamma}{4}\hat{b}^{\dagger}\hat{b}(\hat{b}^{\dagger}\hat{b}+1) + A_{\mathrm{D}}(\hat{b}+\hat{b}^{\dagger})\cos\omega_{\mathrm{D}}. \tag{5.3}$$

We now transform the Hamiltonian into a frame rotating with the drive frequency ω_{D} by applying the unitary transformation

$$U(t) = \mathrm{e}^{\mathrm{i}\omega_{\mathrm{D}}\hat{b}^{\dagger}\hat{b}t},$$

resulting in

$$\begin{aligned} H_{\mathrm{R}}/\hbar &= U^{\dagger}(t)HU(t)/\hbar + \mathrm{i}U^{\dagger}(t)\dot{U}(t) \\ &\approx (\omega_{\mathrm{q}}-\omega_{\mathrm{D}})\hat{b}^{\dagger}\hat{b} - \frac{\gamma}{4}\hat{b}^{\dagger}\hat{b}(\hat{b}^{\dagger}\hat{b}+1) + \frac{A_{\mathrm{D}}}{2}(\hat{b}+\hat{b}^{\dagger}), \end{aligned}$$

where fast rotating terms again have been neglected in a rotating wave approximation. Before we can treat this Hamiltonian by perturbation theory, we have to apply a displacement operator $D_\alpha \equiv e^{\alpha\hat{b}^\dagger - \alpha^*\hat{b}}$, with $D_\alpha^\dagger \hat{b} D_\alpha = \hat{b} + \alpha$ in order to make the drive term small compared to the other terms. This results in

$$\begin{aligned}\tilde{H}_\mathrm{R}/\hbar &= D_\alpha^\dagger H_\mathrm{R} D_\alpha/\hbar \\ &= \left(\omega_\mathrm{q} - \omega_\mathrm{D} - \frac{\Upsilon}{4} - \Upsilon|\alpha|^2\right)\hat{b}^\dagger\hat{b} - \frac{\Upsilon}{4}(\hat{b}^\dagger\hat{b})^2 + (\zeta^*\hat{b} + \hat{b}^\dagger\zeta) \\ &\quad -\frac{\Upsilon}{2}\left(\alpha^*\hat{b}^\dagger\hat{b}\hat{b} + \alpha\hat{b}^\dagger\hat{b}^\dagger\hat{b}\right) - \frac{\Upsilon}{4}(\alpha^2\hat{b}^{\dagger 2} + \alpha^{*2}\hat{b}^2) + \text{const},\end{aligned}$$

where the parameter ζ was introduced as

$$\zeta \equiv \alpha\left[\omega_\mathrm{q} - \omega_\mathrm{D} - \frac{\Upsilon}{2}(1 + |\alpha|^2)\right] + \frac{A_\mathrm{D}}{2}. \tag{5.4}$$

We can now chose the parameter α in the displacement operator such that we get $\zeta = 0$, which eventually cancels the terms linear in $\hat{b}$ and $\hat{b}^\dagger$ in Eq. (5.4). With this choice, the Hamiltonian reads

$$\tilde{H}_\mathrm{R}/\hbar = \underbrace{\left(\omega_\mathrm{q} - \omega_\mathrm{D} - \frac{\Upsilon}{4} - \Upsilon|\alpha|^2\right)\hat{b}^\dagger\hat{b} - \frac{\Upsilon}{4}(\hat{b}^\dagger\hat{b})^2}_{\tilde{H}_\mathrm{R}^0/\hbar} \underbrace{-\frac{\Upsilon}{4}(\alpha^2\hat{b}^{\dagger 2} + \alpha^{*2}\hat{b}^2) - \frac{\Upsilon}{2}\left(\alpha^*\hat{b}^\dagger\hat{b}^2 + \alpha\hat{b}^{\dagger 2}\hat{b}\right)}_{\tilde{H}_\mathrm{R}^1/\hbar}, \tag{5.5}$$

where we already divided the Hamiltonian into the unperturbed part $\tilde{H}_\mathrm{R}^0$, whose eigenenergies $\tilde{E}_k^0$ can be easily calculated, and the perturbation $\tilde{H}_\mathrm{R}^1$, which we will treat with perturbation theory.

Evaluating the individual contributions, a closer analysis shows that higher orders of perturbation theory have to be included for strong drive amplitudes to get better results. Therefore, we evaluated the calculations up to fourth order, which can be found in Appendix A.

$$\begin{aligned}\tilde{E}_k &\approx \tilde{E}_k^0 + \tilde{E}_k^1 + \tilde{E}_k^2 + \tilde{E}_k^3 + \tilde{E}_k^4 \\ \tilde{E}_k^1 &= \langle k|\,\tilde{H}_\mathrm{R}^1\,|k\rangle = 0 \\ \tilde{E}_k^2 &= \sum_{m\neq k}\frac{\left|\langle m|\,\tilde{H}_\mathrm{R}^1\,|k\rangle\right|^2}{\tilde{E}_k^0 - \tilde{E}_m^0} \\ &\vdots \quad \vdots\end{aligned} \tag{5.6}$$

The full expansion of these products would become lengthy and gives no additional insight. It is therefore not presented here. For the later presented analytical data however, the whole expression was included in the calculation. In order to understand the physical meaning of this findings, we restrict our analysis to corrections up to first order. We get

$$\tilde{E}_k^0/\hbar = \left(\omega_\mathrm{q} - \omega_\mathrm{D} - \frac{\Upsilon}{4} - \Upsilon|\alpha|^2\right)k - \frac{\Upsilon}{4}k^2, \tag{5.7}$$

as the first order corrections vanish with $\tilde{E}_k^1 = \langle k| \tilde{H}_\mathrm{R}^1 |k\rangle = 0$. If we recall that we set $\zeta = 0$ after Eq. (5.4), we find for the parameter α:

$$\gamma\alpha|\alpha|^2 + 2\alpha\Delta_\mathrm{D} - A_\mathrm{D} = 0, \tag{5.8}$$

where we introduced the detuning between the drive and the first qubit transition $\Delta_\mathrm{D} = \omega_\mathrm{D} - \omega_\mathrm{q} + \gamma/2$. This gives in general three solutions for α. Since we limit our studies to the non-resonant case, it makes sense to restrict the drive frequencies to a regime where it cannot excite level transitions. When the higher level transitions are below the fundamental transition in frequency ($\gamma > 0$), we want the drive to be above the first qubit transition ($\Delta_\mathrm{D} > 0$) and vice versa. It therefore makes sense to study the case $\Delta_\mathrm{D}/\gamma > 0$, which is true for the later presented measurements. We find that α has only one real solution in this regime:

$$\alpha = \frac{2^{1/3}\left(\sqrt{81A_\mathrm{D}^2\gamma + 96\Delta_\mathrm{D}^3} + 9A_\mathrm{D}\sqrt{\gamma}\right)^{2/3} - 4\sqrt[3]{3}\Delta_\mathrm{D}}{\left(36\left[3\gamma^3\left(27A_\mathrm{D}^2\gamma + 32\Delta_\mathrm{D}^3\right)\right]^{1/2} + 324A_\mathrm{D}\gamma^2\right)^{1/3}}. \tag{5.9}$$

For the other case, where $\Delta_\mathrm{D}/\gamma < 0$, the number of real solutions for α depends on the drive strength A_D. Below a critical amplitude $A_\mathrm{D,crit}$, three real solutions exist, where two of them vanish above $A_\mathrm{D,crit}$. This case is however unpractical for sensing applications, as already mentioned.

With this theoretical analysis, we can make a sanity check by examining Eq. (5.9) in the case of very large detuning, $\Delta_\mathrm{D} \gg A_\mathrm{D}, \gamma$, which eventually corresponds to the dispersive limit previously regarded in Eq. (2.11). In this case, we get $\alpha \approx A_\mathrm{D}/2\Delta_\mathrm{D}$ which implies that α is small and we can neglect higher order terms in perturbation theory. With this and Eq. (5.7), we get in the laboratory frame

$$E_k^0/\hbar = \omega_\mathrm{q}k - \frac{\gamma}{4}k(k+1) - \gamma\left|\frac{A_\mathrm{D}}{2\Delta_\mathrm{D}}\right|^2 k. \tag{5.10}$$

Comparing this with the case of the AC Stark shift caused by a resonator coupled to the qubit, as discussed in Sec. 5.2.1, we find again that the shift of the qubit frequency is proportional to the resonator population by the relation $\langle\hat{a}^\dagger\hat{a}\rangle \propto P_\mathrm{D} \propto {A_\mathrm{D}}^2$, where P_D is the drive power. From this and the result of Eq. (5.1), that is $\Delta\omega_\mathrm{q} = \langle\hat{a}^\dagger\hat{a}\rangle g^2/\Delta_\mathrm{r}$, we additionally see that the drive amplitude of a coupled resonator is given by ${A_\mathrm{D}}^2 = 4g^2\langle\hat{a}^\dagger\hat{a}\rangle$ in the limit of large detuning.

A second conclusion can be drawn from the unperturbed eigenenergies given in Eq. (5.7), where we see that α only acts on terms with k and not with k^2. This means that all levels shift in parallel, as long as α is small and the higher order perturbation terms can be neglected. As said, this especially holds true for the dispersive limit.

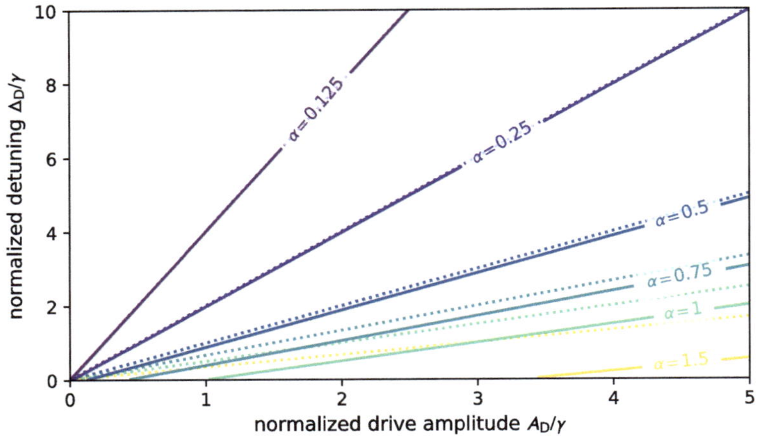

Figure 5.2: Isolines of the displacement parameter α for different combinations of detuning and amplitude. The solution of Eq. (5.9) is calculated for different combinations of drive amplitude A_D and detuning between drive and qubit frequency Δ_D (solid lines). By taking the parameters relative to the anharmonicity γ, this parameter can be eliminated. The dashed lines represent the approximate solution $\alpha \approx A_D/2\Delta_D$, which renders a valid approach for large Δ_D. This plot also demonstrates the validity of the assumption that α is small for $\Delta_D \gtrsim A_D$.

To show the range and validity of these approximations, α was calculated for different values of A_D and Δ_D, and is depicted in Fig. 5.2 as contour plot. This shows that α is small for large detunings $\Delta_D \gtrsim A_D$ and in the general case of small α the linear approximation is justified (dashed lines). Apart from these extremum considerations, any further analysis needs to take into account the higher orders of perturbation theory and we have to calculate the full solution of Eq. (5.6). For a better comparability with the later acquired experimental data, we calculated the energy of the multi-photon transitions from the ground state $|0\rangle$ in the laboratory frame and displayed them for different values of detuning Δ_D and anharmonicity γ in Fig. 5.3 against the drive amplitude. For this graph, a base qubit frequency of $\omega_q = 5.0\,\text{GHz}$ was assumed, similar to the later experiments. The solid lines in Fig. 5.3 a) and b) correspond to a small detuning Δ_D, while the dashed lines represent the analytic data for an increased Δ_D. This demonstrates clearly that the bending of the individual qudit transitions does not only depend on $A_D/2\Delta_D$, as suggested by Eq. (5.10), but on both values independently.

With these findings, the groundwork for using an anharmonic circuit as sensor for detuned microwave signals is set, giving us the opportunity to detect the amplitude and frequency of the signal by measuring the shift of the qudit's transition frequencies. A comparatively small anharmonicity γ, which is usually considered to be unwanted and a drawback of devices like the transmon, can be advantageous for this sensing scheme: As α depends on the relations A_D/γ and Δ_D/γ, the quantity $1/\gamma$ can

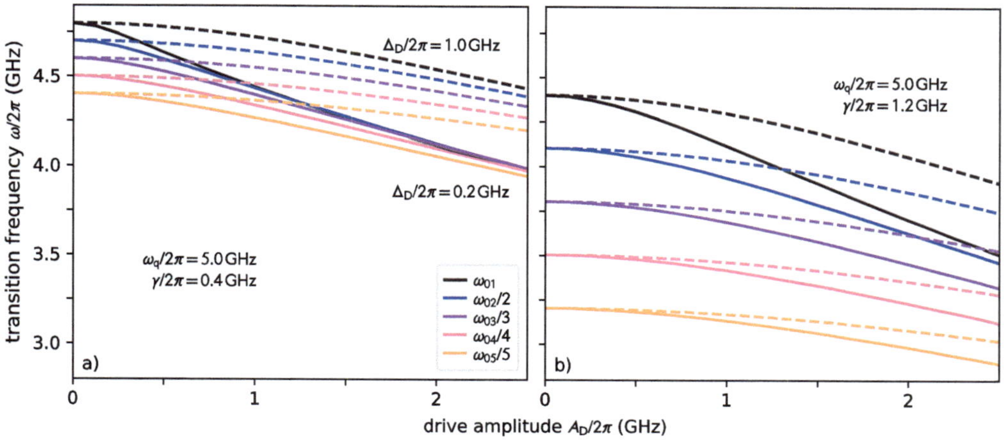

Figure 5.3: Analytically calculated level shifts. The multi-photon transition frequencies are calculated using the analytic formula with perturbation theory up to fourth order. It can clearly be seen that for an increased detuning (dashed lines) the level shift is less pronounced and rather parallel, whereas for a small detuning (solid lines) the levels closer to the drive shift more strongly and eventually cross. Part a) represents a weakly anharmonic circuit with an anharmonicity of $\gamma = 400\,\text{MHz}$, comparable to a standard transmon qudit, whereas part b) demonstrates that the sensing scheme is in principle also applicable to other qudit systems with higher anharmonicity. Adapted from [71].

be seen as an amplification factor for α and hence increases the sensitivity of the sensor. This also becomes clear when comparing Figs. 5.3 a) and b), where γ is increased by a factor of 3 from a) to b).

5.3.2 Numerical solution

In the previous analytical treatment of the driven qudit system, several approximations have been applied, leaving room for an inadequate description of our experimental system. To get a more exact simulation of our physical system, we can perform a numerical analysis, which can also be executed for more complex systems that are hard to describe analytically.

To resolve the level transitions of our anharmonic system under a drive and to do the simulations in closest analogy to the experimental conditions, we simulate not only the influence of the drive tone, but also of the weaker probe tone which we use to excite the level transitions. The simulation Hamiltonian then reads:

$$H_{\text{num}} = \sum_i E_i \left|i\right\rangle \left\langle i\right| + A_{\text{D}}\hbar(\hat{\beta} + \hat{\beta}^{\dagger}) \cos \omega_{\text{D}} t + A_{\text{P}}\hbar(\hat{\beta} + \hat{\beta}^{\dagger}) \cos \omega_{\text{P}} t,$$

where A_{P} and ω_{P} are the probe tone amplitude and frequency, respectively. In contrast to the previously used anharmonic approximation for the qudit, we now leave the individual level energies E_i completely unrestricted and take the anharmonic raising and lowering operators $\hat{\beta}^\dagger$ and $\hat{\beta}$ instead of the harmonic ones. To calculate the anharmonic operators, we have to consider the cross-coupling-constants $g_{ij} \propto \langle i|\hat{N}|j\rangle$ in the basis of the initial transmon Hamiltonian Eq. (2.26), which create additional off-diagonal terms in the operators $\hat{\beta} = \sum_{k>l} g_{kl} |k\rangle \langle l|$ [17].

One limitation of numeric simulations is that the Hilbert space of the problem has to be represented by a matrix. This means that only a limited set of degrees of freedom can be represented in the simulation and every additional degree strongly increases the computation time. To this end, the Hilbert space of the charge operator $\hat{N}$ was restricted to $(\langle\hat{N}\rangle = -15\ldots+15)$ and hence 31 levels, and the transmon excitations are considered up to the $|9\rangle$ level, as higher levels are beyond the Josephson potential barrier.

Master Equation and Lindblad Formalism

To calculate the time evolution of the system described by this Hamiltonian, we use the master equation in the Lindblad form, which describes the evolution of a mixed state with density matrix $\hat{\rho}$ by [155, 156]:

$$\frac{\mathrm{d}\hat{\rho}}{\mathrm{d}t} = -\frac{\mathrm{i}}{\hbar}\left[\hat{H}_{\mathrm{num}},\hat{\rho}\right] + \sum_{\mu}\gamma_\mu\left(\hat{L}_\mu\hat{\rho}\hat{L}_\mu^\dagger - \frac{1}{2}\left\{\hat{L}_\mu^\dagger\hat{L}_\mu,\hat{\rho}\right\}\right), \tag{5.11}$$

with $\{a,b\} = ab + ba$ being the anticommutator. The Lindblad collapse operators $\hat{L}_\mu$ represent the different channels μ through which decoherence acts on the quantum system at a rate γ_μ. Relevant for our analysis are here the qudit decay operator $\hat{L}_1 = \hat{\beta}$ with the energy relaxation rate $\gamma_1 = \Gamma_1$ and the dephasing operator $\hat{L}_2 = \hat{\beta}^\dagger\hat{\beta}/\sqrt{2}$ with a rate of $\gamma_2 = \Gamma_\varphi$.

The third process we can take into account is the spontaneous excitation by thermal energy with the operator $\hat{L}_3 = \hat{\beta}^\dagger$ and a rate of $\gamma_3 = \Gamma_1\bar{n}_{\mathrm{therm}}$, where $\bar{n}_{\mathrm{therm}}$ is the equilibrium thermal population of the system. If we take into account a nonzero thermal population, we also have to correct the decay rate to $\gamma_1 = \Gamma_1(\bar{n}_{\mathrm{therm}} + 1)$ [71]. For our simulations, we assumed a thermal population of $\bar{n}_{\mathrm{therm}} = 0.1$, corresponding to an electronic chip temperature of $T = 100\,\mathrm{mK}$. This not only accounts for possible dissipation caused by the strong drive tone but is also in agreement with experimental findings from other groups [157].

Simulation software and environment

Knowing all input parameters of Eq. (5.11), we can numerically calculate the time evolution of a starting state $\hat{\rho}$ iteratively for small time steps $\mathrm{d}t$. After a steady state is reached, we calculate the expectation value of the number operator $\langle\hat{\beta}^\dagger\hat{\beta}\rangle = \sum_k k\,|k\rangle\,\langle k|$ for the final state. This corresponds to our experimentally observed quantity, the shift of the readout resonator's frequency.

To calculate the time evolution, the software package QuTiP [158, 159] was used, providing a Quantum Toolbox in Python. This toolbox comes, amongst other things, with predefined functions and operators to easily create a representation of pure states, density matrices, operators, and more and allows to calculate expectation values and matrix elements in a very accessible manner. It also features a master equation solver which takes $\hat{H}_{\text{num}}$, $\hat{\rho}_0$, $\hat{L}_\mu$ and a list of time steps $\mathrm{d}t$ as inputs and calculates the evolution of $\hat{\rho}$ for these time steps.

For each combination of A_D, ω_D and ω_P, a full time evolution of the master equation has to be calculated. This is a computation intensive task for a detailed set of combinations, even with the subsequently presented method to decrease the computation time. As the time evolution depends strongly on the choice of parameters, no intermediate results can be exploited for other calculations and each combination has to be calculated independently. On the other hand, this means that the computation task can be massively parallelized, and all combinations could be computed at the same time – if the computation power and memory is available.

To this end, we made use of BwUniCluster in the framework of BwHPC, the high performance computing cluster of the state of Baden-Württemberg. This cluster features 512 computing nodes[3], each equipped with 16 cores operating at 2.6 GHz and a main memory of 64 GB. Being a scientific computation facility, the operating system of the cluster is already equipped with QuTiP, providing easy access to simulations. Since the parallel computations do not need to be done simultaneously nor interact with each other, it was possible to split the computation task into several small jobs which can run on spare computing capacities and hence need less time in the job queue.

Coherence and computation time

In the experiments, we do spectroscopic measurements, meaning that we apply all microwave tones for several milliseconds and are certain that a steady state has evolved. The measurement time per parameter combination is here typically given by hardware limitations and the data acquisition time. For the simulations however, every time interval $\mathrm{d}t$ by which we prolong our analysis takes additional computation time, meaning that we try to reach the steady state as fast as possible.

[3] In total, more than 890 computation nodes contribute to the cluster. The additional nodes are however specialized to certain kinds of jobs and therefore not considered here.

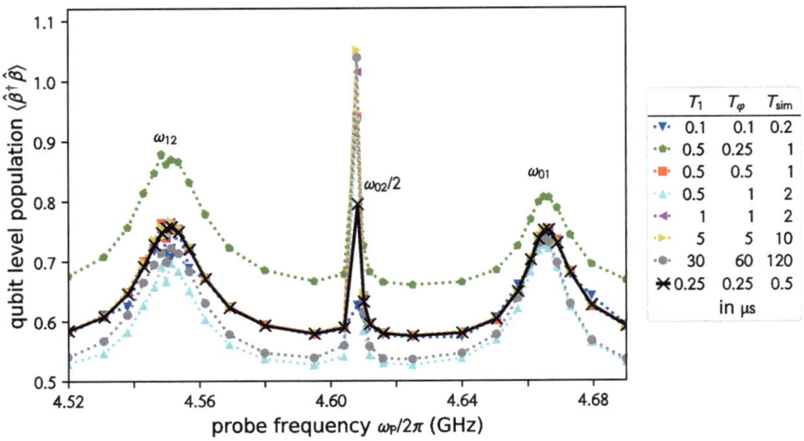

Figure 5.4: Numerically calculated qudit spectrum for different collapse rates. Simulation of the qudit spectrum with a drive amplitude $A_D/2\pi = 0.3\,\text{GHz}$ at $\omega_D/2\pi = 4.95\,\text{GHz}$. To speed up the convergence of the master equation simulation, the decay and dephasing rates $\Gamma_1 = 1/T_1$ and $\Gamma_\varphi = 1/T_\varphi$ are modified, resulting in a shorter overall simulation time T_{sim} until the solution was in a steady state. As a good compromise between simulation time and visibility of all levels, the last combination (black crosses) with $T_1 = 0.25\,\mu\text{s}$, $T_\varphi = 0.25\,\mu\text{s}$ and $T_{\text{sim}} = 0.5\,\mu\text{s}$ was chosen for all other simulations. The lines are guide to the eye and do not represent simulated data. Adapted from [71].

The steady state under a drive is brought about by the Lindblad collapse operators and the corresponding γ_μ rates. For an all-spectroscopic steady state analysis it hence suggests itself to chose higher Γ_1, Γ_φ rates for the simulation compared to the measured rates in order to reduce the overall simulation time.

To verify that this modification does not influence the statement of our analysis, a spectrum of qudit transitions was simulated at fixed input parameters for different Γ_1, Γ_φ rates and an adapted overall simulation length. The results of this analysis are shown in Fig. 5.4, where for each relaxation rate combination three peaks are visible, corresponding to the single photon ω_{12}, ω_{01} and two photon ω_{02} transition. It is apparent that for greater Γ_φ rates and hence reduced T_φ times the lines get slightly broader, whereas a very sharp and high peak can be observed for the longest dephasing and simulation time (gray). The key point is however that the position of the peaks and their general shape does not depend on the chosen rates, leading to the conclusion that it is a valid approach to increase the rates for a spectroscopic analysis and balance between computation time and feature visibility.

To cancel out small oscillations in $\langle\hat{\beta}^\dagger\hat{\beta}\rangle$ which are caused by the drive even in a steady state, the result is always averaged over a period of $T_1/4$ at the end of the simulation time.

5.4 Experimental observation of the multi-mode AC Stark shift

For the experimental measurements of the AC Stark effect, the spectroscopy setup with additional drive line was used, as described in Sec. 4.2.1. To be able to study the influence of the drive also for stronger drive amplitudes, a high power amplifier was used after the drive microwave source. Since the output power of the amplifier is limited, the amplification ratio decreases at high powers. In a separate measurement, we calibrated the output power of the amplifier depending on the power setting of the microwave source to be able to correct our data. In the following, all specified drive powers correspond to the power at the output of the amplifier.

As qudit sample, the single-junction transmon chip with capacitance islands made from TiN was used, described in Sec. 4.3. There, we found a fundamental qubit frequency of $\omega_{01} = 4.755\,\mathrm{GHz}$ together with an anharmonicity $\alpha_1 = -\gamma/2 = -190\,\mathrm{MHz}$ from spectroscopic measurements.

To see the effect of a drive tone on multiple qudit transition frequencies, we apply a drive tone at a fixed frequency of $\omega_{\mathrm{D}} = 4.95\,\mathrm{GHz}$ and increasing values for the power and hence the amplitude A_{D}. This results in a detuning of $\Delta_{\mathrm{D}} = 195\,\mathrm{MHz}$. To probe the transition frequencies, we apply an additional probe tone with a power P_{P}, strong enough to excite multi-photon level transitions, and scan its frequency to excite different transitions. The result is then measured by a weak readout tone applied to the resonator frequency ω_{R} with a power in the single photon regime.

The dispersive shift and with that the qudit population is depicted in Fig. 5.5 a) together with the analytical solutions from Eq. (5.6). We see a good agreement for the first three level transitions of the qudit (solid lines) and get data as expected from Fig. 5.3, where the ω_{01} and $\omega_{02}/2$ transitions approach each other.

However, a variety of additional lines appears in the spectrum, which require further argumentation. These lines can be either explained by a thermal population of the $|1\rangle$ level of the qudit, resulting in the ω_{12} and $\omega_{23}/2$ transitions (dashed lines). Or we have to take into account a multi photon process with virtual energy levels, where the two photons originate not only from the probe tone, but at least one of them is supplied by the drive. This means we can drive the $|0\rangle \rightarrow |2\rangle$ transition with the two photons $\omega_{\mathrm{D}} + \omega_{\mathrm{P}} = \omega_{02}$ (blue dotted line). Due to the fact that the drive frequency is fixed, this curve has a steeper slope compared to the $\omega_{02}/2$ transition (blue solid line). With combinations of multiple photons from drive and probe tone, we are eventually able to describe all observed lines in the spectrum (see legend).

Although the measured spectrum appears to be well represented by the analytic formula, we see deviations starting for $A_D > 0.4\,\text{GHz}$, corresponding to $A_D > \gamma$ and hence an increased parameter α in our analytical treatment. We note here that the proportionality constant in the relation $A_D \propto P_D$ cannot be found experimentally (besides with the AC Stark effect) and therefore was chosen to give a best fit between experiment and numerical simulations, where A_D is directly provided.

To see whether these deviations come from the approximations made for the analytic solution or from experimental imperfections, like wrongly calibrated amplifiers or other elements in the setup that start to get nonlinear at high powers, we simulated the experiment with numerical methods as described before. The results of these measurements are given in Fig. 5.5 b), again with the analytically calculated lines for comparison.

Comparing the numerical data to the experimentally found values, we see a perfect agreement, representing all lines found in the experiment and following the same deviations from the analytic solution. We can therefore conclude that our experimental data are correct and the deviations are caused by the approximations in the analytical solution.

In the numerical data, we see additional transitions, which are not visible experimentally. These transitions include the higher excited $|4\rangle$ and $|5\rangle$ states, meaning that the amplitude of the probe tone was chosen too high compared to the experiment. We also see a general increase in background qudit population with increasing drive amplitude, suggesting that the drive starts to populate the qudit. This background was calibrated away for the experimental data to increase the visibility of the features.

5.5 Sensing scheme

After having demonstrated that the experimental outcome matches to the analytical calculation and even better to the numerical simulation, we are able to invert the scheme and use the qudit as a sensor for off-resonantly applied signals by measuring the shift of the qudit transition frequencies and deducing the parameters leading to this shift. This means we can measure the amplitude of any drive field that couples to the qudit by means of the qudit itself, leading to a new in-situ measurement method. In doing so, the qudit can be under the exact same conditions as it will be used for other experiments later.

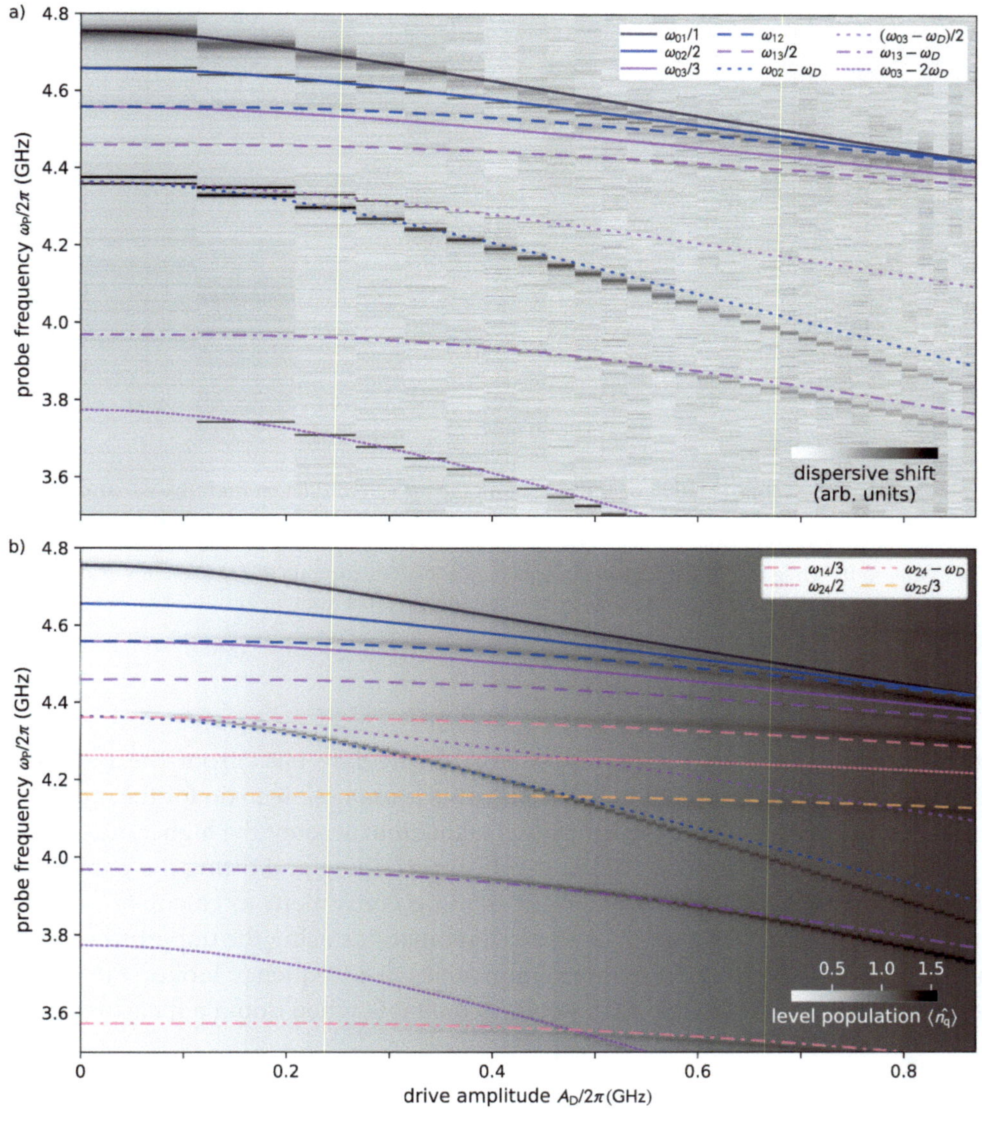

Figure 5.5: Measured and numerically simulated AC Stark shift together with analytical transitions. a) The measured transmon level transitions for a drive at $\omega_D = 4.95\,\text{GHz}$ and increasing drive amplitude A_D. The analytic solution of Eq. (5.6) is displayed as colored lines. Especially for the level transitions (solid lines), a good fit to the measured data is observed. Other transitions including thermally excited states or drive photons can also be modeled by the analytic solution (dashed/dotted lines). Deviations for higher drive amplitudes are explained by the use of a simplified transmon Hamiltonian in the deduction of the analytic solution. The data is column-wise normalized as the drive also influences the resonator frequency. b) Numeric simulations in close analogy to the experiment. The data is well comparable to the measured data and shows the same deviations from the analytical solution, since the full transmon Hamiltonian is used for the simulation. Due to a slightly higher probe power in the simulation compared to the experiment, more transitions are visible which include the $|4\rangle$ state of the qudit. Adapted from [71].

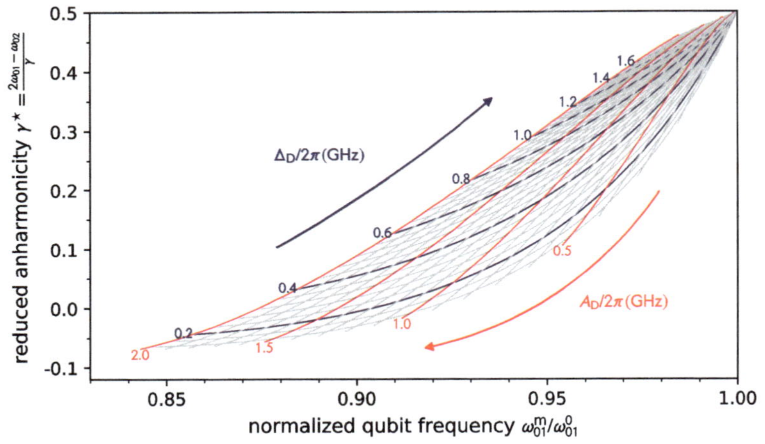

Figure 5.6: Graphical representation of the FLT. From the measured data on qudit base frequency and anharmonicity, each in reference to the unperturbed qudit parameters, we can pick a point on the frequency lookup table (FLT) and resolve the combination of drive amplitude A_D and frequency ω_D which leads to this shift. While this FLT was calculated using the analytic formula in Eq. (5.6), it is also possible to fill the table with data from numeric simulations, resulting in more accurate results. Adapted from [71].

5.5.1 Frequency lookup table

To operate the qudit as a sensor, we have to calculate back from the measured frequency shift to the originating drive field. This is impossible to do for the numerical simulations and increasingly difficult when taking into account the higher order terms in perturbation theory. Instead of iteratively approaching and fitting the parameters for each time we do a measurement, it is more convenient to calculate once the transition frequencies for a set of different combinations using the unperturbed qudit parameters. We store these transition frequencies in a frequency lookup table (FLT) and only have to find the closest matches in the table when doing a measurement.

For a graphical representation of the FLT, we chose to plot the table contents against a reduced anharmonicity $\gamma^\star = (2\omega_{01} - \omega_{02})\gamma$ and the normalized qubit frequency $\omega_{01}^{m}/\omega_{01}^{0}$, where ω_{01}^{m} is the measured qudit frequency and ω_{01}^{0} is the unperturbed frequency. Plotting into this coordinate system lines of constant Δ_D and A_D, respectively, brings us to a representation like in Fig. 5.6. With a measured $\gamma^\star$ and $\omega_{01}^{m}/\omega_{01}^{0}$, we can then directly read the values of Δ_D and A_D responsible for this shift.

5.5.2 Measuring the transmission function

Putting the sensing scheme to the test, we use it to measure the transmission function $T(\omega)$ of our setup. This includes the whole way from the microwave source via the cryostat wiring to the sample box and even inside the box from the transmission line to the qudit. Being able to measure only parts of it when warmed up to room temperature, the whole transmission function is usually completely inaccessible. It has to be noted here that the proportionality constant between drive amplitude at the chip and applied power at room temperature is now frequency dependent with $A_\mathrm{D} \propto T(\omega_\mathrm{D})P_\mathrm{D}$.

For the measurement, we fix the applied drive power P_D and scan the drive frequency ω_D in a second experiment. Scanning the probe tone, we can identify the qudit transitions and hence calculate $\gamma^\star$ and ω_{01}^m, which we use to look up the drive amplitude A_D. Since the frequency ω_D of the drive is known, we can either search for the best matching drive amplitude for the parameter combination $(\gamma^\star,\ \omega_{01}^\mathrm{m},\ \omega_\mathrm{D})$ or we leave the drive frequency as open parameter as well. It is of course expected that the results for the case of a given ω_D are more accurate, but the sensing scheme should nevertheless be able to determine the frequency of the drive tone. For increasing detuning, the influence of the AC Stark shift decreases, resulting in a decreasing resolution and accuracy.

The results of this measurement are given in Fig. 5.7, where the raw experimental data, i.e., the dispersive shift at different probe powers and drive frequencies is given in subfigure c). The precision of the measurement is indicated by the shaded green and gray areas in Fig. 5.7 a) for the amplitude and Fig. 5.7 b) for the drive, which visualize the spread in results when varying the fitted ω_{01} and $\omega_{02}/2$ values by $\Delta\omega/2\pi = \pm 1\,\mathrm{MHz}$. This is a worst case estimate based on the rather coarse step size when scanning the probe tone, $\Delta\omega_\mathrm{P}/2\pi = 1\,\mathrm{MHz}$. For the results, this gives on average $\Delta A_\mathrm{D} = \pm 92\,\mathrm{MHz}$ and $\Delta\omega_\mathrm{D} = \pm 124\,\mathrm{MHz}$, if ω_D is taken as a free parameter. As the sensitivity for the drive frequency and therefore the resolution in this parameter gets very poor for a detuning of more than 5 GHz, no errorbars are shown in this regime in Fig. 5.7 b) and the extracted points are plotted in gray. For the case of a fixed ω_D, the error in amplitude reduces to $\Delta A_\mathrm{D} = \pm 12.5\,\mathrm{MHz}$.

This error estimation is very crude and also the measurements themselves have to be seen more as a proof-of-principle demonstration than as a highly optimized measurement. The resolution could be greatly improved by different means which will be discussed in the next section. Nevertheless, the principal shape of $A_\mathrm{D}(\omega_\mathrm{D})$ makes clear that this method can be used to make statements about the transmission function of the cryostat. For an ideal setup at constant input power, we would expect an increase of A_D close to the qudit transition frequencies due to the frequency-dependent dipole moment of the qudit and for higher frequencies a monotonous decrease converging

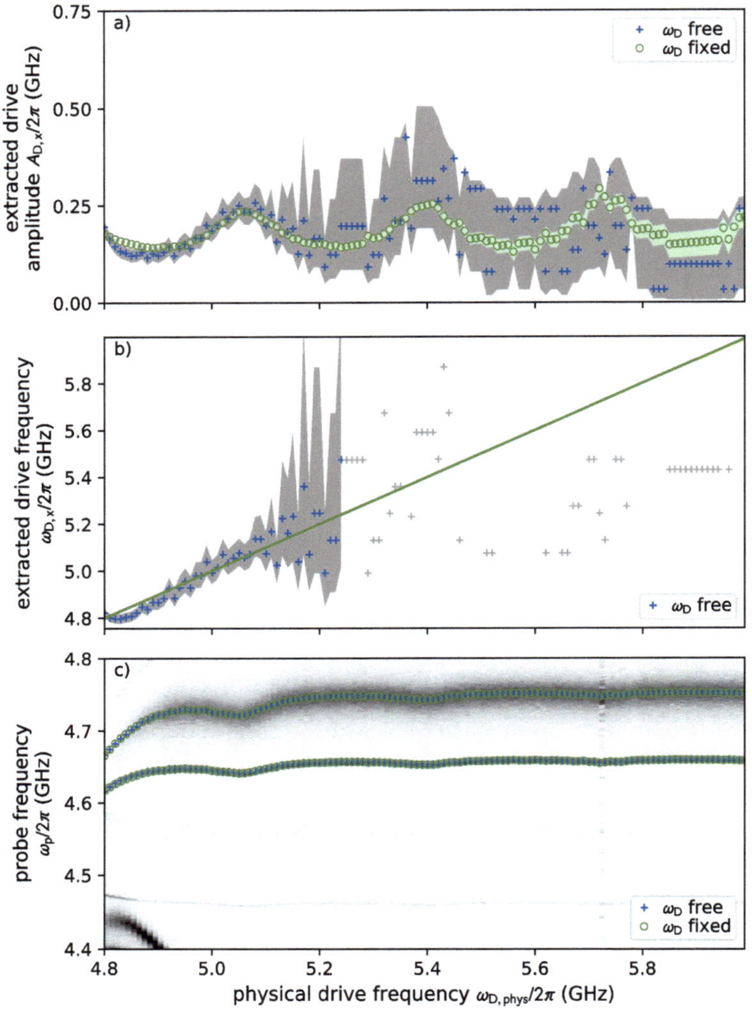

Figure 5.7: Proof-of-principle demonstration of the sensing scheme. For a constant drive power P_D at room temperature, the drive frequency is scanned and the qudit transition frequencies are recorded. With the help of a FLT, we can calculate back to the amplitude a) and frequency b) of the drive. As explained in the main text, the accuracy of the amplitude detection can be greatly improved by taking the known drive frequency as additional input (green circles). For the blue crosses, both A_D and ω_D are free. The shaded areas illustrate the uncertainty of the sensor when taking into account an accuracy of ± 1 MHz for both $\omega_{01}/2\pi$ and $\omega_{02}/4\pi$. For more than 0.5 GHz detuning between drive and first qudit transition, the frequency determination fails in this demonstration, therefore the data points are grayed out and no error bars are shown. Subfigure c) shows the measurement data, i.e., the dispersive shift of the resonator in gray scale. To verify the validity of taking ω_D as input parameter, the transition frequencies are calculated from the data shown in a) and b) and plotted as blue crosses (green circles) when having ω_D as free output parameter (fixed input parameter). For both methods, we find a good agreement to the measured data within the resolution of the measured data. Adapted from [71].

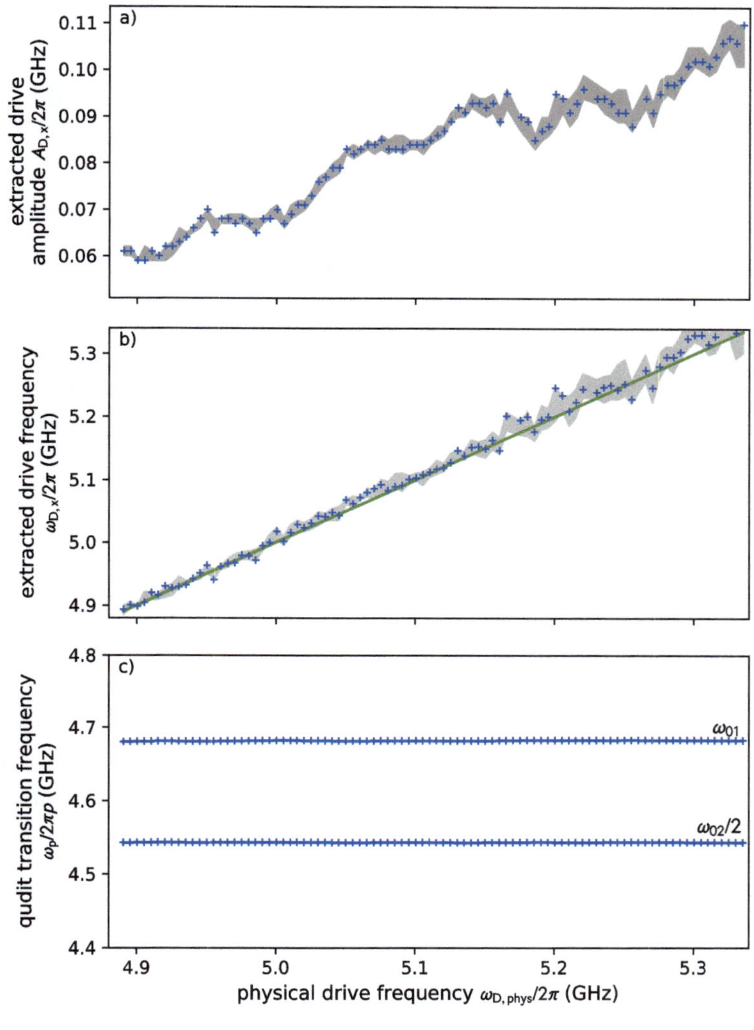

Figure 5.8: Demonstration of increased sensitivity and resolution. By taking Ramsey measurements instead of spectroscopic measurements as in Fig. 5.7, the resolution in measuring the qudit transition frequencies is improved, leading to a greatly increased accuracy of the sensor. In addition, the FLT was generated by numeric simulations and not by the analytical approximation as before. These measurements are taken on a different chip and in a different cryostat than the ones in Fig. 5.7, so the values are not directly comparable. Also, due to the fact that no input amplifier was used for the drive, the drive amplitude is much lower. With reduced drive power and increased minimal detuning, the shifts in ω_{01} and $\omega_{02}/2$ are much smaller and barely visible in subfigure c). The overall accuracy in both amplitude and frequency is however greatly increased, visualized by the much smaller gray band around the data points. Courtesy of M. Kristen [72, 160].

towards a constant value. The measurement however only shows the onset of this increase towards ω_{01} and oscillations of A_{D} for higher detuning. These oscillations can be explained by a non-ideal transmission of the microwave lines through the cryostat, being soldered and connected at each temperature stage of the cryostat.

5.6 Further advances

After the publication of the paper associated with this chapter [71], demonstrating a proof-of-principle experiment, the method was further studied and improved, in particular by the associated Master's thesis of Maximilian Kristen [160] which also resulted in a joint paper [72]. The results of this work are briefly summed up in the following.

5.6.1 Drawbacks of approximate theory

To increase the accuracy of the sensor, it is important to create a reliable FLT. With the deviations seen in Fig. 5.7 it is reasonable to assume that the analytic solution is not perfectly suitable for the calculation of a good FLT. The origin of these deviations was found by performing numerical simulations with the approximated transmon Hamiltonian Eq. (5.3), resulting in a very good agreement between the analytic solution and the simulation result. This leads to the conclusion that the approximation of the transmon Hamiltonian is too inaccurate and dominates the discrepancies to the experimental results, whereas the fourth order perturbation theory can be seen as a valid approximation.

To this end, the FLT was calculated by numerical simulations, taking more computation time but finally giving more accurate sensing results.

5.6.2 Time-domain approach

As an alternative to spectroscopic measurements which require fine-grained frequency sweeps, pulsed measurements can be used. Under certain conditions they allow for a very fast determination of the qubit transition frequency by utilizing an optimized detection scheme [161]. This scheme is based on Ramsey oscillations explained in Sec. 4.2.3 and can use advanced methods to minimize the number of time steps Δt necessary to get a value for ω_{01}^{m}. The same technique can be applied to measure Ramsey oscillations between the $|1\rangle$ and $|2\rangle$ states and hence infer the ω_{12} frequency.

These measurements have been performed on a different qudit sample in a different cryostat, so the results, shown in Fig. 5.8, are not directly comparable to the ones previously found. However, it is obvious that the sensing accuracy is now greatly improved based on an enhanced FLT and finer resolved measurement data. With the improved FLT, the data for the extracted drive frequency now matches the applied drive frequency much better, especially in the regime of low Δ_D, where the analytically calculated FLT has major inadequacies. With an average error of $\Delta\omega_{01} = \pm 6.7\,\text{kHz}$ and $\Delta\omega_{12} = \pm 10.5\,\text{kHz}$, the accuracy of the input data was also improved by about two orders of magnitude compared to the first demonstration experiment.

5.7 Conclusion

In summary, we have demonstrated a sensing scheme for microwave tones by a superconducting anharmonic multi-level system, based on the AC Stark effect. This sensing scheme can be of great use for the optimization and engineering of quantum circuits as it enables to measure the cross-talk of microwave structures to a qudit on the exact same chip as intended for other quantum experiments. Using directly the qudit as sensor eliminates the need to fabricate additional chips for pre-characterization of spurious couplings and helps to validate the results of microwave simulations.

With the possibility to measure and characterize oscillating signals in frequency and amplitude, this sensor also lends itself to the field of fundamental research, where it can be used to detect resonant frequencies of other systems in a quantum environment. As semiconducting structures typically stop working at cryogenic temperatures and dissipate a lot of heat, many commonly used and commercially available sensors for microwaves are not usable at millikelvin temperatures. A simultaneous detection of amplitude and frequency exploiting the characteristic changes in the qudit's transition frequency fingerprint can be seen as clear advantage, which outperforms other qubit-based sensing schemes.

For all kinds of cryogenic experiments involving microwave signals, this sensor can be used to measure the frequency-dependent microwave transmission through the cryostat and the quantum chip, which is otherwise not accessible at the operation conditions. We are able to characterize the transmission for a broad frequency range of more than 1 GHz above the qudit transition frequency, where the resolution starts to decrease with increasing detuning.

For the sensing scheme, technologically simple spectroscopic measurements are sufficient, but pulsed measurements can be used to increase resolution and decrease the measurement time. By performing Ramsey experiments, we can exploit the high sensitivity of the qudit phase towards changes in the transition frequency. With this, the sensor accuracy is greatly enhanced and we demonstrated an average amplitude resolution of $\Delta A/2\pi = 3.4\,\mathrm{MHz}$ as well as an average frequency uncertainty of $\Delta\omega/2\pi = 25\,\mathrm{MHz}$ [72]. The accessible sensing range and resolution can be adapted in time-domain measurements by adjusting the reference frequency until one of the following limits is reached: For low signal amplitude and large detuning, the frequency shifts decrease until the phase accumulated in a Ramsey measurement within the decay time of the qudit states is no longer measurable. Typically, the lifetime of the $|2\rangle$ state is the limiting factor here, as the shift of this transition is smaller and the lifetime of this state is shorter than for the $|1\rangle$ state. In the other limit of low detuning and strong drives, resonant effects like level transitions and the Autler-Townes splitting have to be taken into account, which can finally disturb the sensing protocol.

Finally, our theoretical description can be seen as an extension of the Jaynes-Cummings-Hamiltonian for multi-level systems which is not restricted to the dispersive regime but can be applied also for low detunings.

6 Superconducting quantum bits in magnetic fields

Using superconducting transmon qubits for hybrid systems and as sensors in magnetic applications requires a proper qubit function under the influence of magnetic fields. In the following chapter, I present our investigations on a transmon qubit in a magnetic field to fathom these limits of application. After an introduction covering existing research in this field, I will present spectroscopic measurements, which reveal additional properties of the qubit. To characterize the losses of oscillating superconducting systems, the quality factor of the resonator is studied, directly followed by time domain measurements on the qubit. Here, different loss mechanisms are explained and evaluated, before the chapter concludes with measurements on the pure dephasing rate. The findings presented in this chapter lead to the publication of Ref. [70].

6.1 Introduction

To use a superconducting qubit for quantum sensing applications and for hybrid systems, it is of great use to know its limits of application. Especially the critical parameters of superconductivity are here of interest, as they disturb the qubit's working scheme. These are in particular the critical current, critical temperature and the critical magnetic field.

Limitations in terms of current can be seen by either inducing a constant current through a SQUID loop, as it was shown for the tunable sample, or by creating an AC electric field across the JJ, which was in depth explained in the last chapter in terms of the AC Stark shift. Both methods lead to a controllable decrease in qubit transition frequency, which has been studied and used in application [17, 71].

The behavior of a qubit at elevated temperatures is known to be limited by a strongly increased number of quasi-particles (QPs) and long before the eventual breakdown of superconductivity [162]. Both relaxation and coherence times have been observed to rapidly decrease above a temperature of 100 mK [163, 164].

Besides that, the influence of high frequency phonons has been studied by the application of ultrasonic pulses to superconducting quantum bits [165], as well as the dependence on stray infrared light [166].

For magnetic fields however, no in-depth study was made which investigates the coherence properties of a superconducting qubit under increased magnetic fields, until the coherence breaks down. So far, only the influence of a magnetic field up to the order of 100 µT was studied, where a slight improvement of the coherence properties was found due to the creation and pinning of flux vortices which act as QP traps [113]. Other research groups assume superconducting qubits to break down at very little fields, although they do not provide any measured data to support this claim. Instead, they propose other, more stable junctions, like the proximitized semiconducting nanowire [167].

The general consent is however to screen magnetic fields as good as possible and meanwhile, a multilayered shielding based on permalloy and superconductors is considered to be state of the art [168, 169].

There are many applications like quantum sensing of magnetic excitations [29], the creation and harnessing of Majorana fermions [170] or research fields like cavity quantum magnonics [64, 66], which all require the qubit to be in or close to magnetic fields. It is therefore inevitable to study the influence of magnetic fields to superconducting qubits. For all these applications, it would be beneficial to maximize the magnetic field in which the qubit can be operated. As we know from Sec. 2.3.2, we should then align the magnetic field in parallel to the chip surface. At this point it should be however noted that a parallel field orientation over the whole chip is hard to achieve, because the structured superconductor on the chip bends the field lines and leads to a flux focusing for the regions free from superconducting material. For specific applications, the inclusion of structured magnetic materials will also heavily influence the local magnetic field distribution. Therefore, no additional effort was spent to assure a parallel alignment but the chip was straightened in the solenoid by eye, as described in Sec. 4.1.3.

6.2 Qubit transition frequency in an in-plane magnetic field

Before we perform pulsed measurements, we firstly have to measure the dependence of the qubit transition frequency on the magnetic field. As described in Sec. 2.4.2, the critical current of a JJ in an in-plane magnetic field periodically goes down, where the shape follows a Fraunhofer pattern. With $\omega_{01} \propto \sqrt{I_c}$, a similar behavior is expected

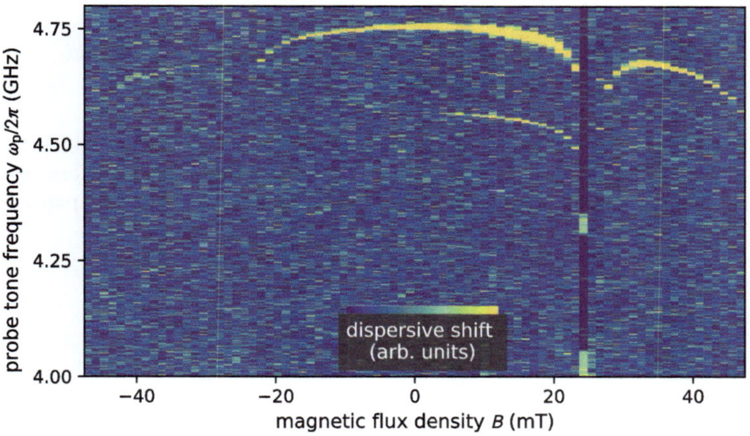

Figure 6.1: Measured transmon frequencies in a magnetic field. The dispersive shift of the resonator is shown color-coded to visualize the transmon transition frequencies under the influence of a magnetic field. For each field value, the readout frequency was adjusted to follow the resonator frequency. The absolute value of the normalized dispersive shift is plotted here to account for these offsets. In the original data, the dispersive shift changes its sign due to a changing quality factor of the resonator, which would distract the colorscale. According to our expectations, we see a decrease in frequency and a periodic revival. Compared to the expected Fraunhofer pattern of a single JJ in a magnetic field however, the top at $B = 0$ is too flat and the side maxima are too high in frequency. Adapted from [70].

is expected when placing the qubit junction in an in-plane magnetic field, such that flux can enter the insulating barrier. This would include a pronounced maximum at Φ/Φ_0 and periodic side maxima with decreasing amplitude.

6.2.1 Measured data

When we place the TiN qubit introduced in Sec. 4.3 inside the solenoid and apply a magnetic field, we observe a periodic decrease in ω_{01} following this behavior, as shown in Fig. 6.1. If we compare this data to the expectation of a Fraunhofer interference pattern, the measurement data shows a flat top at $\Phi = 0$, very steep drops at $\Phi \approx \Phi_0$, and side maxima which are almost as high as the main maximum. The pattern can hence not be explained by a simple rectangular junction cross section. Since the junction was fabricated in a shadow angle evaporation process, deviations from a perfectly rectangular cross section are very likely. Therefore, we Fourier transformed different possible junction geometries but did not find any $I_c(B)$ respectively $\omega_{01}(B)$ relation resembling the measured data. A different process is hence needed to explain the measured results.

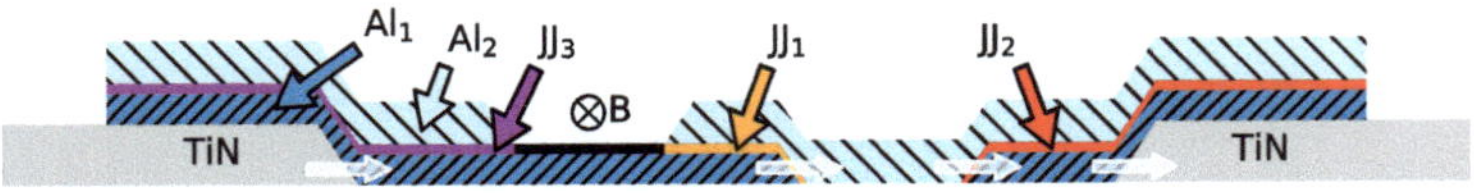

Figure 6.2: Illustration of all created JJs. Following the fabrication scheme in Fig. 4.11, we find two additional JJs being created. While JJ_3 only shunts the lower Al_1 layer and is therefore negligible for low kinetic inductance films, the qubit excitation current flows through JJ_2 which explains the additional contribution. The current path is indicated by the white arrows. Adapted from [70].

6.2.2 Revisiting junction fabrication

If we revisit the junction fabrication process for the TiN sample from Sec. 4.4.1, we see that besides the intended small Josephson junction JJ_1 in the center of the structure, two other junctions, JJ_2 and JJ_3 exist, which are commonly neglected due to their large junction area and hence high critical current. These junctions are illustrated in Fig. 6.2. Due to the multi-step fabrication process in which first the TiN capacitance pads are fabricated and then the junctions on top of this, the current has to flow across JJ_1 and JJ_2 to go from one electrode to the second, as indicated by the white arrows. JJ_3 only shunts the lower Al layer, which is a negligible effect as long as the inductance of this layer is low.

If we qualitatively analyze the influence of the parasitic junction JJ_2, which is in series to the intended qubit junction JJ_1, we see the following behavior: As long as the parasitic junction is not close to being blocked by a flux quantum, its critical current is very high and not limiting or influencing the qubit current The qubit transition frequency can be therefore assumed to be rather flat and unimpressed by the magnetic field. In the case of $\Phi/\Phi_0 \approx n$ ($n \in \mathbb{N} \setminus 0$) for the parasitic junction, we will see a steep decrease of the qubit frequency. This goes along well with our measured data, and we will deduce a quantitative formula for a transmon qubit with two junctions in series in the following.

Transmon Hamiltonian for two JJs in series

To derive the transmon Hamiltonian for two JJs in series, we start with the general Hamiltonian introduced in Sec. 2.5, which has contributions from the charging energy in Eq. (2.24) and the Josephson energy in Eq. (2.25). In the general case, we have:

$$H = \frac{Q^2}{2C} + \frac{\Phi_0}{2\pi} \int I(\varphi) \mathrm{d}\varphi. \tag{6.1}$$

To solve this, we have to find the current-phase-relation $I(\varphi)$ for the total system. We see that the phase acquired by the individual junctions, φ_1 and φ_2, adds up to a total phase, and the current through the junctions has to be the same:

$$\varphi_\Sigma = \varphi_1 + \varphi_2$$
$$I(\varphi_\Sigma) = I_{c,1} \sin \varphi_1 + I_{c,2} \sin \varphi_2.$$

We introduce the ratio $r = I_{c,2}/I_{c,1}$ between the junction's critical currents to facilitate the writing, and chose $r \geq 1$ without loss of generality. This gives

$$\tan \varphi_2 = \frac{\sin \varphi_\Sigma}{r + \cos \varphi_\Sigma}$$

and we get for the overall current-phase relation

$$I(\varphi_\Sigma) = I_{c,2} \sin \left(\arctan \frac{\sin \varphi_\Sigma}{r + \cos \varphi_\Sigma} \right). \tag{6.2}$$

The integral of this equation can be written in a closed form and we find

$$\int I(\varphi_\Sigma) \mathrm{d}\varphi_\Sigma = -I_{c,2} \frac{\sqrt{r^2 + 2r \cos \varphi_\Sigma + 1}}{r} = -I_{c,1} \sqrt{r^2 + 2r \cos \varphi_\Sigma + 1}.$$

This results in a total system Hamiltonian of

$$H = 4E_C \hat{N}^2 - E_{J,1} \sqrt{r^2 + 2r \cos \varphi_\Sigma + 1}, \tag{6.3}$$

where we have again used the charging energy $E_C = e^2/2C$ and the Josephson energy of the qubit junction $E_{J,1} = I_{c,1} \Phi_0 / 2\pi$, and number and phase operators as in Sec. 2.5.

In analogy to Eq. (2.27) we perform a Taylor expansion up to fourth order in φ_Σ and find

$$H \approx 4E_C \hat{N}^2 + E_{J,1} \left(\frac{r}{2r+2} \varphi_\Sigma^2 - \frac{r(r^2 - r + 1)}{24(r+1)^3} \varphi_\Sigma^4 \right) \tag{6.4}$$

$$\text{with } N = i \left(\frac{\hbar^2}{32} \frac{E_{J,1}}{E_C} \frac{r}{r+1} \right)^{1/4} (\hat{b}^\dagger - \hat{b})$$

$$\text{and } \varphi = \left(\hbar^2 \frac{2E_C}{E_{J,1}} \frac{r+1}{r} \right)^{1/4} (\hat{b}^\dagger + \hat{b}).$$

With the same approximations as for the simple transmon case, this gives

$$H = \sqrt{8E_C E_{J,1} \frac{r}{r+1}} \, \hat{b}^\dagger \hat{b} - \frac{r^2 - r + 1}{(r+1)^2} \frac{E_C}{2} \hat{b}^\dagger \hat{b} \, (\hat{b}^\dagger \hat{b} + 1). \tag{6.5}$$

In the limit of $r = I_{c,2}/I_{c,1} \to \infty$, where JJ_1 is dominating (i.e., limiting) the qubit's behavior, this formula recovers the simple approximated transmon Hamiltonian from

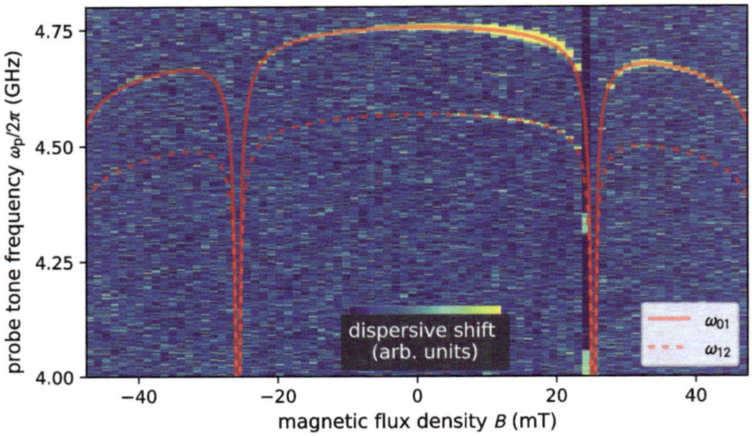

Figure 6.3: Two-junction transmon. The analytic transition frequencies for a transmon qubit with two JJs in series are shown on top of the measured data. Here, JJ_1 gives the overall envelope and JJ_2 creates the periodic minima. For the measured data, the same normalization was applied as in Fig. 6.1. The analytic formula matches the measured data very well for ω_{01} and ω_{12}, which appears in the measurement due to thermal population of the $|1\rangle$ state. Adapted from [70].

Eq. (2.29). We also see from this formula that the anharmonicity changes with $\frac{r^2-r+1}{(r+1)^2}$ which means that the effective anharmonicity of a transmon with two junctions in series is always lower than if only one junction was present. For the limit $r = 1$, where the two junctions in series have the same critical current, the anharmonicity is reduced by 75 %.

When we assume an almost rectangular cross-section for the two junctions, each critical current should follow $I_c = I_c^0 \left|\operatorname{sinc}\frac{B-B_\Delta}{B_{\Phi_0}}\right|$. Besides the periodicity field B_{Φ_0} we also introduced an offset field B_Δ here, which accounts for stray background magnetic fields. Fitting Eq. (6.5) to our data, we find a good agreement for ω_{01} and ω_{12}, shown by the solid and dashed red lines in Fig. 6.3. The extracted values for each junction are given in Table 6.1.

From the periodicity of the pattern, we can calculate back to the junction's size with $\Phi_0 = B_{\Phi_0} A$. The effective cross section of the junction is given by $A = (d + 2\lambda_L)l$ with l being the junction length. Since the magnetic field reaches into the superconductor

Table 6.1: Parameters of the two Josephson junctions. B_Δ is the offset field, B_{Φ_0} the periodicity, and l the length of the junction.

	E_J/h	B_Δ	B_{Φ_0}	l
JJ_1	16.15 GHz	1.8 mT	300 mT	209 nm
JJ_2	300 GHz	-0.2 mT	25.5 mT	2.46 µm

on a length scale of λ_L above and below the junction, this has to be added to the thickness d, which is the oxide barrier. For our junctions, we take $d = 1\,\mathrm{nm}$ and $\lambda_L = 16\,\mathrm{nm}$ for aluminum. With this we get the junctions' lengths to be $l_1 = 209\,\mathrm{nm}$ and $l_2 = 2.46\,\mu\mathrm{m}$, which agrees well with the design parameters.

The increased sensitivity against magnetic fields and the abrupt drops in the qubit transition spectrum can be disadvantageous for many applications. To avoid this spurious influence, the parasitic junctions JJ_2 and JJ_3 can be shunted with a bandage in an additional fabrication step [171]. Alternatively, other fabrication methods apart from shadow angle evaporation can be used, e.g., overlap junctions [172]. In this process, no additional junctions are created.

Due to the fact that we cannot see the first minimum of the qubit junction in our measured data, not to mention further maxima, the extracted size of the qubit junction has to be taken carefully. What has been neglected in the treatment up to now is that the critical current of a Josephson junction depends also on the superconducting gap of the leads, which in turn depends on the magnetic field. This is modeled by the Ambegaokar-Baratoff relation [173, 174]:

$$I_c R_n = \frac{\pi}{2e} \Delta(B) \tanh \frac{\Delta(B)}{2 k_B T}. \tag{6.6}$$

For our values of Δ and T, the tanh term is negligible and we see that $I_c \propto \Delta$. Together with $\Delta(B) = \Delta_0 \sqrt{1 - (B/B_c)^2}$, where B_c is the critical field, we get a relation for $I_{c,1}(B)$ and can calculate the qubit transitions. Assuming a critical field of $B_c = 168\,\mathrm{mT}$ which gives the best fit to our data, we see no difference to the previously presented model within the range of our acquired data. This can be seen by the good match of the colored dashed lines in Fig. 6.4. In this limit, we assumed that there is no Fraunhofer-type reduction of $I_{c,1}$, i.e., the qubit junction is assumed to be point-like. In reality, both effects coexist at the same time.

6.3 Resonator behavior in magnetic fields

Before we come to time resolved data on the qubit coherence, we shortly take a look at a similar measure: the quality factor of the readout resonator. Since the transmon can be explained as a modified resonator, it seems likely that the transmon is affected by the same spurious mechanisms as the resonator and the resonator's quality factor $Q = \omega/\Delta\omega$ evolves similar to the qubit decoherence time with $T_2 = 2/\Delta\omega_q$, where $\Delta\omega_q$ is the qubit linewidth. At the same time, the resonator acts as readout device and eventually limits the signal-to-noise ratio of our measurement within reasonable

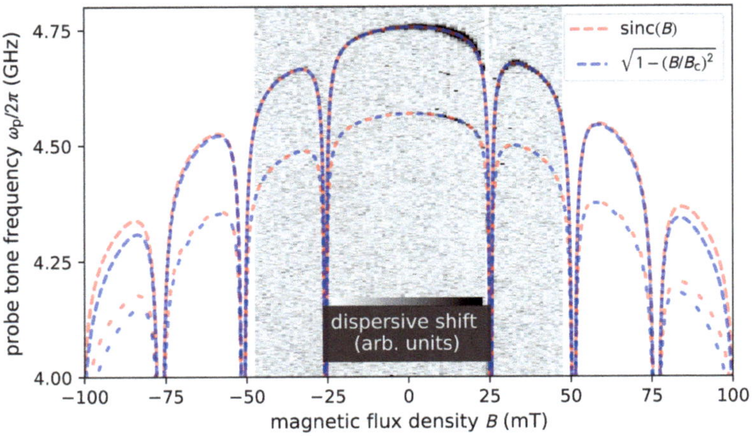

Figure 6.4: Comparison between different $I_c(B)$ relations for JJ_1. The critical current of JJ_1 creates the envelope for the transition frequencies. It can be either suppressed by a Fraunhofer-type interference pattern as explained in Sec. 2.4.2 (red dashed line) or by a reduction of the superconducting gap energy Δ (blue dashed line). The narrow-dashed lines show the ω_{01} transitions, and the wide-dashed lines show ω_{12}. For the measured data in the background, the same normalizations apply as before. Within the range of acquired data, no difference between the to methods can be seen. Increasing the measurement range is also not possible with the used setup due to the strong reduction of the readout resonator's quality factor and therefore decreased signal-to-noise ratio. Adapted from [70].

acquisition time. This is already visible in Fig. 6.1, where the qubit signal fades out for high field values due to the decreasing quality factor of the resonator.

Superconducting resonators in magnetic fields have already been studied in detail, with a focus on an alignment of the magnetic field perpendicular to the resonator [130, 131, 175], QP recombination after infrared pulses [116] or a comparison between zero-field-cooled and field-cooled data [115].

Our measured data are presented in Fig. 6.5 and match well to the previously published data. The data are extracted by means of a circle fit [136], where applicable. For very low quality factors, the circle fit did not converge due to a small phase contribution of cable resonances which could not be calibrated away. For these measurements, a Lorentzian fit to the amplitude was used to obtain the resonance frequency, which is shown as open symbols in Fig. 6.5.

The data correspond well to the in-plane data previously measured by other groups [131], where they compare it to a loss rate calculated from the classical Bean model [132, 133]. This simulation matches our data well for the case of a weakly inhomogeneous RF current distribution. Although Fig. 6.5 c) suggests that the resonator is completely interspersed with flux vortices at $B \approx 100\,\mathrm{mT}$, a closer look in the data does not

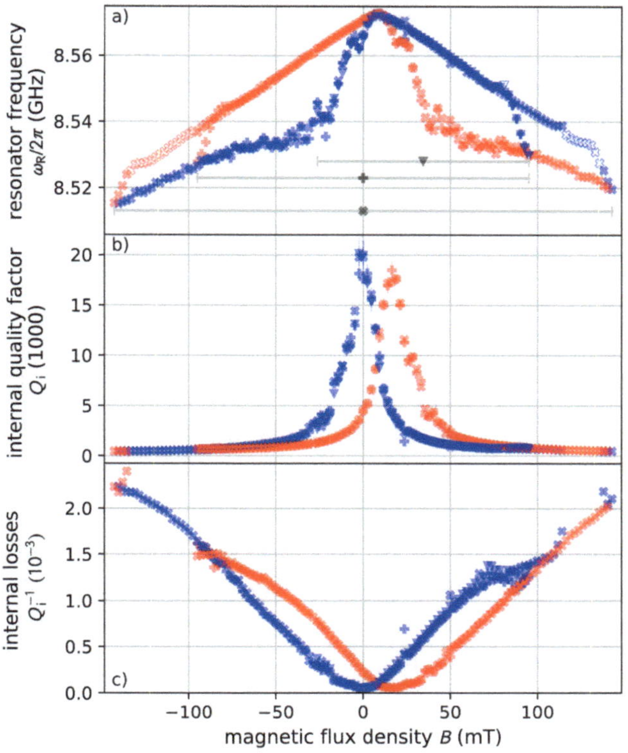

Figure 6.5: Resonator data in a magnetic field. Spectroscopic data of a TiN $\lambda/2$ resonator in microstrip geometry. As errorbars, the uncertainty of the fit values is indicated. A decrease in resonance frequency a) can be seen as well as a strong decrease of its internal quality factor b). Both parameters are subject to a strong hysteresis. Red (blue) symbols are used for an upsweep (downsweep) of the magnetic field. Filled symbols represent data extracted from a circle fit. Where this fit did not converge, a Lorentzian fit was applied (open symbols) to extract the resonance frequency. The gray bars and symbols denote the field range of the corresponding sweep data. c) shows the inverse of the quality factor for a better comparison with the data presented in Ref. [131]. Adapted from [70].

support this statement, since the data extracted by a Lorentzian still shows a difference between up and down sweep for the loaded quality factor (data not shown).

For the coupling quality factor Q_c we see no significant change for different values of B, which was expected since this quantity is defined purely by the geometric coupling of transmission line and resonator. We find an average value of $Q_c = (9.3 \pm 1.3)\,\mathrm{MHz}$.

From the resonator measurements we already see an offset field of $B_{\mathrm{offs}} = 8.5\,\mathrm{mT}$ in both $\omega_R(B)$ and $Q_i(B)$, which we will find again for the later measurements of the qubit coherence parameters. We assume that this offset field appears due to the presence of stray fields from magnetized components around the qubit chip, which are partially aligned perpendicular to the chip. If we assume small misalignment between the chip and the coil field, we find that an applied field of 8.5 mT would compensate a perpendicular magnetic field of $B \approx 450\,\mu\mathrm{T}$, which is well on the order of typical stray fields, for a misalignment of about $\alpha \approx 3°$.

6.4 Qubit coherence in magnetic fields

6.4.1 Measurement scheme

To acquire reliable data and account for the variety of changes which appear when changing the magnetic field, a measurement scheme was constructed which ramps the current of the solenoid to the specified field value and then recalibrates the necessary experimental parameters to start the measurement, as follows:

1. The readout tone is scanned around the previously measured frequency of the resonator. The recorded amplitude signal is fitted to a Lorentzian and the resonance frequency ω_R taken as new readout frequency to get a good contrast in the dispersive measurements.

2. The manipulation tone is scanned across the previous qubit transition frequency to find the new qubit frequency ω_{01} by again fitting a Lorentzian to the phase data of the recorded dispersive shift.

3. Applying a Rabi sequence at the previously determined ω_{01} frequency and fitting the data to an exponentially decaying cosine yields the time t_π for a π pulse.

4. A sequence of T_1, T_2^R and/or T_2^E is executed to get the desired coherence parameters.

In the steps 1-3, selection mechanisms apply to sort out non-converging fits by detecting too high errors or fit results that leave a predefined range. The sequence is then repeated once for the same field value and skipped if it fails the second time. The number of averages as well as points in frequency (steps 1 and 2) or time (steps 3 and 4) is reduced to be able to perform the whole sequence within 10 minutes while still maintaining a data quality good enough for the fitting, despite the decreased SNR.

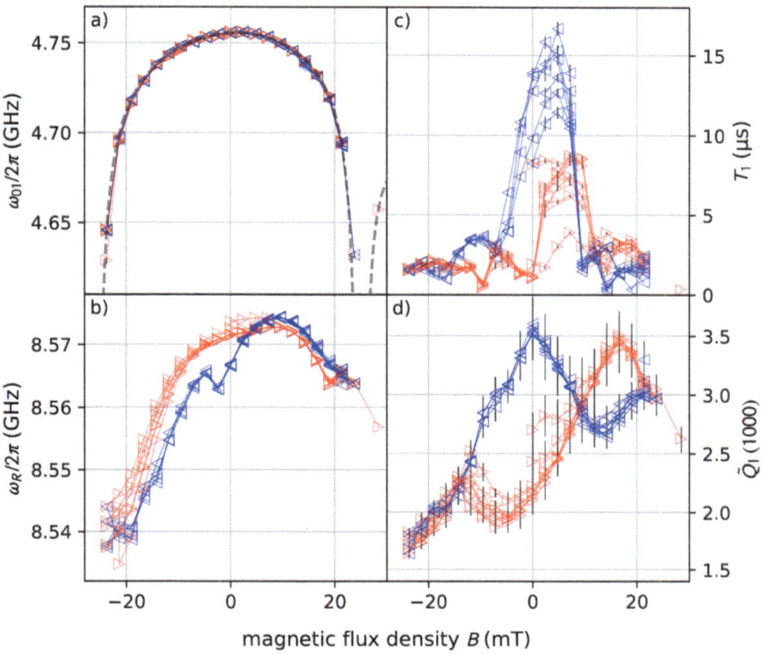

Figure 6.6: Time-domain measurements of a transmon in a magnetic field. Red (blue) triangles show data acquired on an upsweep (downsweep) of the magnetic field. The colored lines are only guide to the eye to trace related data points. As errorbars, the standard deviation of the fits is indicated. a) The measured qubit transition frequency matches well the calculated transition frequency (gray dashed line) and shows no hysteresis. b) The resonator frequency shows in contrast a clear hysteresis and a small decrease with applied field. c) A hysteresis is also seen for the qubit decay time T_1, which shows a pronounced maximum at low fields. d) The resonator's quality factor extracted from time domain measurements can only be seen as an indicative value due to the non-linear amplitude detection scheme in the time domain setup. Properly measured quality factors are found in Fig. 6.5. Adapted from [70].

6.4.2 Measured data

The acquired data for multiple measurement sequences in the range of $B = \pm 23.7\,\mathrm{mT}$ is shown in Fig. 6.6. Again, all data points acquired on an upsweep of the magnetic field are marked in blue, where the data from a downsweep are shown in red. In Fig. 6.6 a) we see again the measured qubit transition frequency, which follows Eq. (6.5) and shows no hysteresis. This means that the qubit transition frequency is dominated by the characteristic Fraunhofer pattern of the two serial junctions.

6.4.3 Separation of loss mechanisms

Looking at the measured data for the relaxation time T_1, presented in Fig. 6.6, we see a prononounced maximum at low fields and a distinct hysteresis between up and down sweep. We can now analyze this data phenomenologically by separating the losses of the system as

$$\frac{1}{T_1} = \Gamma_1 = \Gamma_{\text{hyst}} + \Gamma_{\text{non-hyst}} + \Gamma_{\text{const}}.$$

In this classification, Γ_{hyst} accounts for loss mechanisms showing a hysteretic field dependence, $\Gamma_{\text{non-hyst}}$ collects losses that depend directly on the magnetic field strength, and the losses associated with Γ_{const} do not depend on the magnetic field.

As already discussed in Sec. 2.6, a variety of different loss mechanisms comes into play, especially when applying a magnetic field. For Γ_{hyst}, we assume that the main contribution is coming from the creation, annihilation and movement of flux vortices. As described in Sec. 2.6.4, it is known both theoretically and experimentally that the vortex configuration in a superconductor exhibits a large hysteresis on its magnetic history, which can be explained by the Bean model [132, 133]. Due to the large aspect ratio of the qubit island with a diameter of 554 µm and a thickness of 40 nm, we assume that the small misalignment to the external field suffices in order to generate flux vortices perpendicular to the film. This also seems to be a realistic assumption when regarding other experiments, where much effort is required to get a field alignment without the creation of perpendicular vortices. Otherwise, these vortices dominate the loss and frequency properties of the resonator [176].

From a comparison to the resonator loss data we see that shapes and signs of $\tilde{Q}_l$ and T_1 are generally similar (compare Figs. 6.6 c) and d)), but the qubit loss data feature a much more pronounced overall envelope. We attribute this envelope – and therefore a contribution of $\Gamma_{\text{non-hyst}}$ – to the presence of excitations in the superconductor, i.e., quasiparticles (QPs). From the discussion in Sec. 2.6.3 we know that the losses due to QPs are proportional to the density of QPs and hence to the squared magnetic field, $\Gamma_{\text{QP}} \propto x_{\text{QP}} \propto B_{\parallel}^2$. The QP density is not reported to have a hysteretic dependence on the effective field and the relaxation to an equilibrium QP density is expected to happen within a few µs. The hysteretic vortex configuration however affects the effective field in the superconductor as well as the QP creation and recombination rate, and with that the QP density.

Other losses like radiative decay and losses to two-level-systems or defects in the dielectrics, which are discussed in Sec. 2.6.2, are assumed to be independent from the magnetic field and are therefore summed up in Γ_{const}.

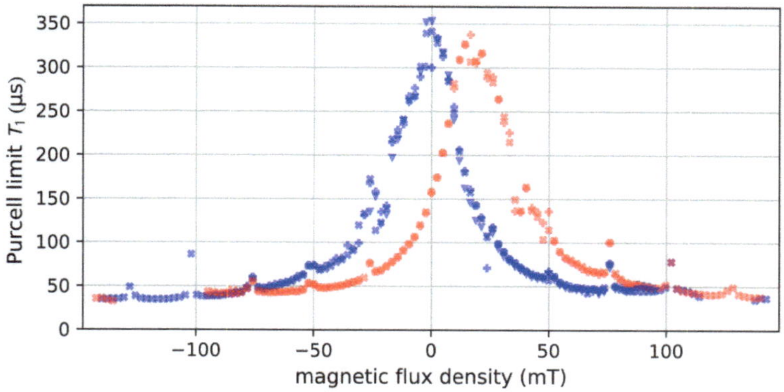

Figure 6.7: Purcell limitation of the qubit decay time. Calculated Purcell contribution to the T_1 time of the qubit, following Eq. (2.32) using the measured resonator data shown in Fig. 6.5 and the calculated qubit transition frequency for the corresponding field values. Regular spikes in the calculated T_1 time can be seen, coming from a strongly decreased qubit transition frequency due to the blockage of JJ_2 and therefore an increased detuning between resonator and qubit. The resulting T_1 times are at no point limiting the qubit coherence and the Purcell losses can therefore be neglected or considered to be constant against the magnetic field for this sample. Note the larger range of the x axis compared to the T_1 data presented in Fig. 6.6.

For the Purcell loss, as introduced in Sec. 2.6.1, a variety of strongly changing parameters have to be taken into account. These are for instance qubit and resonator frequency and resonator linewidth. The Purcell loss is therefore also subject to changes in the magnetic field. Taking the spectroscopic data as presented in Fig. 6.5 and combining it with the analytic transmon transition formula from Eq. (6.5), we find the Purcell loss contribution depicted in Fig. 6.7. Since the Purcell contribution ranges from $\Gamma_{\text{Purcell}} \approx 350\,\text{Hz}$ to $\Gamma_{\text{Purcell}} \approx 40\,\text{Hz}$ for fields much stronger than the ones in the previously presented measurements, we conclude that the Purcell contribution does not change the qubit's relaxation rate significantly for any applied magnetic field value.

Summing up the argumentation in this section, we see that we can model the qubit relaxation rate by $\Gamma = \Gamma_{\text{const}} + C(B - B_{\text{offs}})^2$ [113, 115, 116]. If we fit this envelope to our measured data, we find $\Gamma_{\text{const}} = 53.4\,\text{kHz}$, $C = 0.785\,\text{kHz}\,\text{mT}^{-2}$ and $B_{\text{offs}} = 2.25\,\text{mT}$. The fit is modified here such that only solutions which are below the measured data are taken into account. We assume the remaining hysteretic losses to come from the entering and movement of flux vortices. The different offset field B_{offs} compared to the previously stated offset is explained by the fact that in this evaluation, the data at high magnetic field and therefore low decay time are weighted more strongly compared to the previous case, where the maximum of T_1 was evaluated. Although the parabola shown in Fig. 6.8 is a proper envelope for the measured data, we do not want to make any claim that this is a proof for our chosen partitioning of the loss mechanisms. In fact, the different loss mechanisms are not distinguishable by

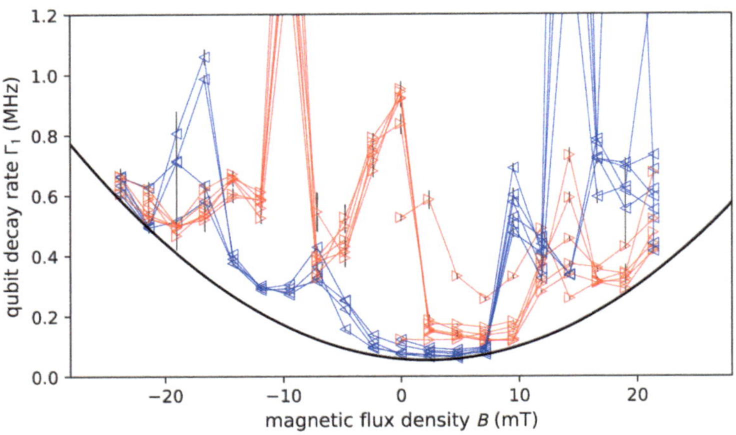

Figure 6.8: Modeling loss contributions. The decay rate $\Gamma_1 = 1/T_1$ is presented, where red (blue) triangles again mark upsweeps (downsweeps). The black line is a lower-limit-fit including a constant relaxation rate and a parabolic contribution as described in the text. The colored lines connect sequentially taken data points and are a guide for the eye. Adapted from [70].

our measurement technique. Our partitioning only represents the most obvious loss channels and their dependence on magnetic fields.

6.4.4 Increased magnetic field

If we further increase the magnetic field, we observe quantum coherence of the qubit up to values of $B_{\mathrm{appl}} \approx 40\,\mathrm{mT}$. This measurement is shown in Fig. 6.9. Although $T_1 = 0.49\,\mu\mathrm{s}$ is significantly reduced at this high field, we clearly observe Rabi oscillations and an exponential decay after a π pulse, as demonstrated in the inset of Fig. 6.9. The quality factor Q_1 of the resonator is significantly reduced at this point, which explains the low SNR. For even higher values of B, we are not able to track the qubit transition due to this SNR and the decreased T_1, T_2 times which result in a broadening of the qubit linewidth.

Performing a down sweep from this field value (blue triangles) does not show a pronounced maximum as before but only a slight increase in T_1 over a broader range. We see however a fine structure in the data, showing multiple drops in T_1 which coincide with the onset of a deviation from the Fraunhofer pattern, followed by a jump in frequency. We attribute this effect to the presence of flux vortices in the qubit islands due to the previously applied high fields. Their local field influences the field seen by the junction and therefore qubit frequency and coherence.

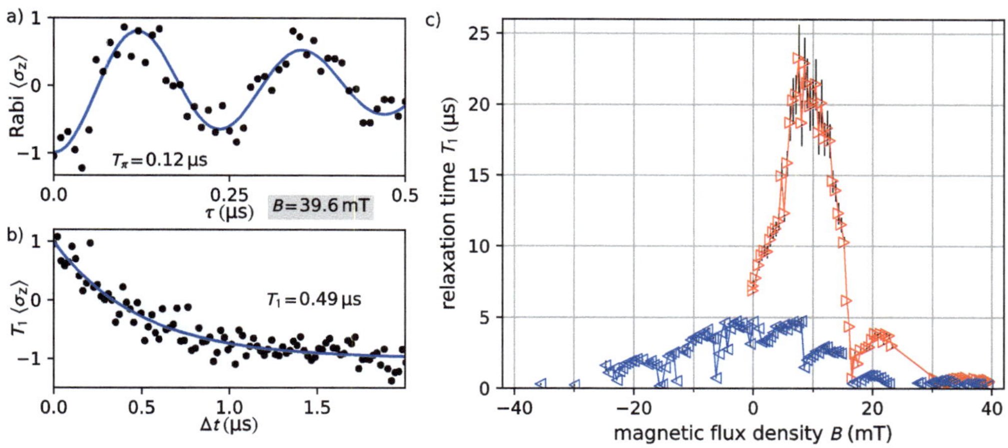

Figure 6.9: Coherence properties up to higher field values. Increasing the magnetic field further, we can see qubit coherence up to $B = 39.6\,\mathrm{mT}$ and a) still observe Rabi oscillations and b) measure a relaxation rate. On a downsweep of the field afterwards (blue triangles in c)), the T_1 time does not recover as before but shows multiple abrupt changes. These jumps are attributed to the movement or annihilation of a flux vortex. The values shown for $\langle\sigma_z\rangle$ in a) and b) are nomalized to the accessible qubit values, i.e., an increased residual population is calibrated away. Adapted from [70].

6.5 Pure dephasing rates

Up to now, we only considered the relaxation rate of the qubit, Γ_1. To get information on the phase coherence of the qubit, which is typically the quantity of interest for quantum computation and many quantum sensors, we simultaneously measure T_1 and T_2^{R} and relate them via

$$\frac{1}{T_2^{\mathrm{R}}} = \Gamma_2^{\mathrm{R}} = \frac{1}{2}\Gamma_1 + \Gamma_\varphi,$$

where Γ_φ is the pure dephasing rate. If we find this dephasing rate to be constant, this means that the qubit behavior is well described by the relaxation measurement alone.

6.5.1 Measurement scheme

In order to get physically connected Γ_1 and Γ_2^{R} rates, we acquire the measurement data for both measurements in turn, such that temporal fluctuations of the qubit properties influence both measurements likewise. This technique was developed for the measurements presented in Ref. [31] and is explained in more detail there. Without doing so, the two rates do not need to be result from the same perturbations and measurements

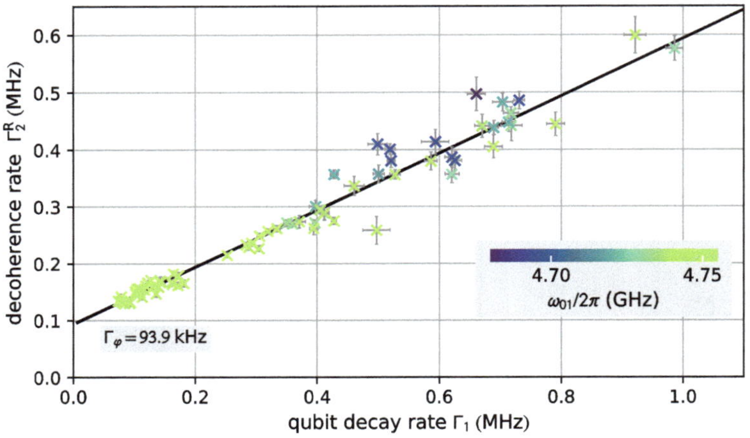

Figure 6.10: Pure dephasing rate. The pure dephasing rate Γ_φ can be seen as the y-intercept in this visualization. For the areas with a higher slope of $\omega_{01}(B)$ (dark crosses) we could expect a higher pure dephasing rate due to a higher sensitivity to magnetic field noise and fluctuations of the solenoid current. A clear trend is however not visible and all data points match well to the linear regression. The standard errors from the fits for T_1 and T_2^R are shown as error bars. Adapted from [70].

with $T_2^R > 2T_1$ can occur, which makes no sense from a physical point of view. Before each measurement, the re-calibration routine as presented in Sec. 6.4.1 is executed.

6.5.2 Measured data

We plot Γ_2^R against the simultaneously acquired Γ_1 rate, as shown in Fig. 6.10, and fit the data to a straight line. From the y-intercept we find a pure dephasing rate of $\Gamma_\varphi = 93.9\,\text{kHz}$. Due to the steeper slope of $\omega_{01}(B)$ for the regions close to the blockage of JJ_2, we would expect an increased Γ_φ for data points with lower ω_{01}. For the measurements presented in Fig. 6.10, where ω_{01} is color-coded, we however see no clear correlation between Γ_φ and ω_{01}.

The causality between noise in the solenoid current S_I and the resulting $\Gamma_{\varphi,I}$ is given by [177]:

$$\Gamma_\varphi^I = \pi \left(\frac{\partial \omega_{01}}{\partial I} \right)_z^2 S_I(\omega \ll k_\mathrm{B}T), \tag{6.7}$$

where the relevant scale for ω is the time between the Ramsey pulses, being on the order of $\omega/2\pi \approx 100\,\text{kHz}$. The measured power spectral density of our current source is about $S_I \approx 10^{-15}\,\text{A}^2\,\text{Hz}^{-1}$, with which we can calculate $\Gamma_\varphi^I(B)$ as shown in the upper part of Fig. 6.11. In the lower part, $\Gamma_\varphi^I(B)$ is color-coded onto the qubit transition

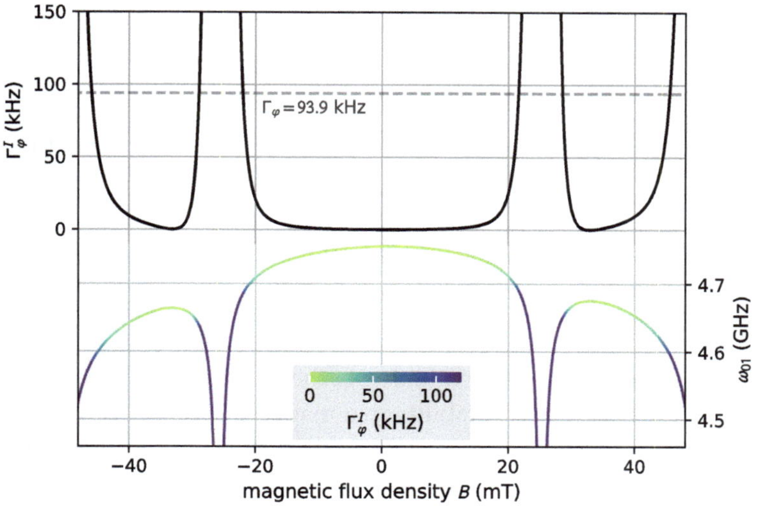

Figure 6.11: Calculated dephasing rate. From the measured noise power spectral density S_I of our current source we can calculate the current noise contribution to the pure dephasing rate and plot it against the applied magnetic field (black curve). The gray dashed line represents the extracted pure dephasing rate from Fig. 6.10. To indicate the areas, where the qubit dephasing rate is limited by the current noise, the calculated Γ_φ^I is color-coded into the qubit transition frequency $\omega_{01}(B)$. This shows that only in the regions of a steep slope of $\omega_{01}(B)$ the current noise exceeds the other pure dephasing channels.

frequency. From this plot, we can see that Γ_φ is mainly not limited by the current noise but by other factors which are constant against the magnetic field. Only in the areas of a steep slope of $\omega_{01}(B)$, the dephasing rate is increased by the magnetic field noise.

6.6 Conclusion

In this chapter, we measured the behavior of a superconducting transmon qubit in a magnetic field, where we demonstrated quantum coherence up to a flux density of $B = 40\,\mathrm{mT}$. We explained the observed decay rates by different mechanisms and saw a strong indication that the loss mechanisms can be split into constant, hysteretic and non-hysteretic losses. Studying the pure dephasing rate, we found it to be constant and not affected by the magnetic field, such that the dephasing is governed by the qubit's relaxation.

From the observation of the qubit's transition frequency, we found the influence of a second, parasitic junction. This junction is typically present in circuits fabricated by shadow angle evaporation, but commonly neglected as it has a high critical current. Due to its large cross section, it has an increased sensitivity to in-plane magnetic fields. To calculate the influence of this parasitic junction, we derived a theoretical description which resulted in an approximated Hamiltonian for a transmon qubit with two serial junctions.

Analog to the qubit, we studied the behavior of a superconducting resonator in a magnetic field. With our results being comparable to previously published data by other groups, we find a strong increase of the losses in the resonator, expressed by an increased linewidth. This broad linewidth finally appears as bottle neck for the measurement, as it impedes a fast and efficient readout of the qubit at even higher field strengths.

From our measurements, and the partitioning of the qubit losses, we can derive multiple handles for increasing the qubit's stability in a magnetic field. Most importantly, traps for moving flux vortices and excess quasiparticle could be implemented in future designs to decrease the influence of a magnetic field on the qubit coherence. Shunting or avoiding the parasitic junction is also strongly recommended, if a monotonous decrease of the qubit in the magnetic field is requested.

Despite these possibilities for optimization, we find the non-optimized version of a state-of-the-art transmon qubit to be well compatible with magnetic fields beyond the critical field of bulk aluminum. Even though the decay and coherence times are strongly suppressed compared to the unperturbed case, they are still on the order of many other superconducting qubit designs, which have been used for outstanding publications. With our findings, the usability range of superconducting qubits as versatile sensors is increased and a promising basis for future developments in superconductor-magnet-hybrid devices is created.

7 Towards coupling of qubits and magnetic materials

Combining superconducting qubits with magnetic oscillations promises versatile applications in quantum sensing and quantum computation, as will be presented in this chapter. So far, we have studied in this work how we can perform quantum sensing with a transmon qubit and how a superconducting qubit behaves in a magnetic field. Following this route, our natural next step is the realization of a coupling between qubits and a magnetic material.

In this chapter, I will start with a description of oscillating magnetic systems and provide basic theoretical background to understand the origin and the mechanisms of these oscillations. We then go into detail on the coupling mechanism between the qubit and this system, and show an experimental approach to realize a direct coupling. Finally, we present first results on this approach, showing a rich set of interesting features which are still subject to further research. This can be seen as a starting point for further experiments with superconducting quantum bits in a magnetic field, coupled to magnetically tunable systems.

7.1 Introduction

After the revolutionary development in the field of electronics in the past 75 years and the emerging number of applications in spintronics, a very active field of fundamental research is constituted by the field of magnonics. Here, the information is not stored and transmitted by the charge or spin of single electrons. Instead, collective excitations of the electronic spin system are used, where the long-range order of a ferro- or ferrimagnetic material supersedes the need for a physical transportation of electrons. These collective excitations can be seen as synchronous precessional motion of the spins around an externally applied field. In a quasi-particle description, they are called magnons and are described in greater detail in the next section.

In the field of magnonics, many basic computation elements and logic gates have been realized and demonstrated, for example a magnonic transistor [52], magnonic

AND and XOR gates as well as directional coupler and half-adder [178]. Due to the decreased propagation velocity of magnons compared to electromagnetic waves, magnonic resonators are much shorter in length and therefore have a smaller footprint. This is especially beneficial for applications where a dense packaging is required. Since the spin wave can propagate without a particle transport, the loss mechanisms known from electronics do not apply.

Although these devices have been actively developed and engineered for several years, little is known about the quantum behavior of magnons. Many experiments up to now have been performed at room temperature and high powers to create a non-equilibrium spin distribution. Magnonic properties at low excitations and fundamental studies about the included damping mechanisms however have only recently become a part of fundamental research. While cavity magnonics already enables to study the magnonic system at low temperatures and excitation powers, single magnon operations can only be performed when including a single photon device, i.e., a qubit. This creates the basis for experiments in quantum magnonics.

Apart from the fundamental interest in exploring the single-excitation behavior of magnons and the underlying effects, one more applied goal is driving the research of qubit magnon coupling: Due to the interaction of magnons with optical photons [46, 179], a bidirectional transducer between superconducting qubits and optical photons may be realized. This would be one of the missing building blocks for a quantum computer in a quantum internet, since superconducting qubits are among the most promising candidates to build the computation cells, and optical photons are almost exclusively used to transfer quantum information between different places.

7.2 Magnons and magnetic oscillations

In ferromagnetic materials, the elementary magnetic moments preferably align along a common axis due to an exchange interaction, which forms a long-range order. Since this interaction is stronger than the repulsion of their dipolar magnetic fields, all spins within a domain spontaneously align in parallel below a critical temperature [61]. For the following discussion, we assume that an external field is applied to create a uniform bias field throughout the sample such that all spins align in the same direction. We also assume that the temperature is low compared to the magnetic field, such that $k_B T < \vec{\mu} \cdot \vec{B}$, where k_B is the Boltzmann constant and $\vec{\mu}$ is the magnetic moment of the individual spins.

One intuitive excitation of the ground state would be to flip exactly one spin in the system, as illustrated in Fig. 7.1 a). This is however energetically not favorable, since

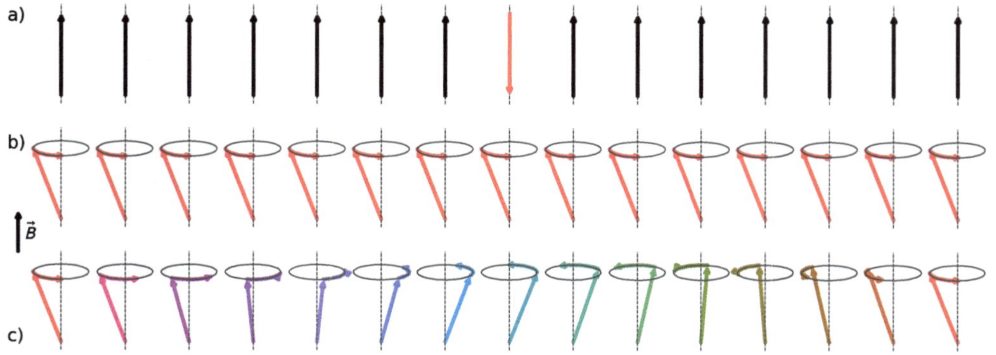

Figure 7.1: Excitations in a ferromagnet. Under the influence of an external field $\vec{B}$, all spins align along the z axis in the ground state. a) One form of excitation is to flip one spin. This is energetically not favorable since the exchange energy between the flipped spin and its neighbors is needed. Energetically more favorable is a small tilt of all spins in the system, which then start a precessional motion around the external field. This can be in-phase for all spins as shown in b), or wave-like as shown in c).

this spin would stand contrary to all its neighbors. An energetically more favorable solution is a distributed excitation, where all spins are slightly tilted, which can be seen as a linear superposition of all possible single-spin flips, illustrated in Fig. 7.1 b).

If we compare the spin in the magnetic field to a spinning gyroscope in gravity, we see that a tilt immediately results in a precessional motion around the direction of the external force to keep the angular momentum constant. This precession is illustrated by the small curved arrows in Fig. 7.1 b) and c). Depending on parameters such as shape, material and the uniformity of the excitation, the precession can be in-phase as in Fig. 7.1 b) or can have a wave vector $\vec{k}$ with $|\vec{k}| = 2\pi/\lambda$ as illustrated in Fig. 7.1 c) for $\lambda = 1$, where λ is the wavelength.

7.2.1 Magnon frequency

Instead of describing the precession of each spin individually, typically the Holstein-Primakoff transformation [180] is applied, which describes all excitations in the ferromagnet by a superposition of spin wave excitations, or magnons. The Hamiltonian for a ferromagnetic spin wave then reads:

$$\hat{H} = \sum_{\vec{k}} \hbar\omega(\vec{k})\hat{m}^{\dagger}_{\vec{k}}\hat{m}_{\vec{k}}, \tag{7.1}$$

where the sum goes over all possible wave vectors $\vec{k}$. In the general case and especially for structured magnetic materials, the magnon dispersion $\omega(\vec{k})$ does not only depend on the absolute value $|\vec{k}|$, but also on the direction of propagation, the sample dimensions, and the orientation of the external magnetic field. Analog to other bosonic systems, $\hat{m}_{\vec{k}}^{\dagger}$ and $\hat{m}_{\vec{k}}$ are the magnon creation and annihilation operators for the corresponding $\vec{k}$ vector.

If we assume that all spins are precessing in phase, $\vec{k} = 0$, we can replace the individual spins by a large macrospin. The precession frequency of this macrospin is given by

$$\omega_{\mathrm{P}} = \left|\gamma \vec{B}\right|,$$

where γ is the gyromagnetic ratio. We know that the flux density $\vec{B}$ inside the magnetic material is given by the external field $\vec{H}_{\mathrm{ext}}$ and the internal magnetization $\vec{M}$, i.e., $\vec{B} = \mu_0 \left(\vec{H}_{\mathrm{ext}} + \vec{M}\right)$. This gives for the resonance frequency

$$\omega_{\mathrm{P}} = \gamma \mu_0 \left|\vec{H}_{\mathrm{ext}} + \vec{M}\right|.$$

In this formula however, the shape of the sample was not included. If we assume a thin film in parallel to the external field, we obtain [181]

$$\omega_{\mathrm{P}} = \gamma \mu_0 \sqrt{H_{\mathrm{ext}} \left(H_{\mathrm{ext}} + M\right)}. \tag{7.2}$$

This equation for the spin-wave frequency ω_{P} only holds for an in-plane orientation, does not cover the influence of a magnetocrystalline anisotropy field and assumes $\vec{k} \approx 0$, i.e., a homogeneous AC excitation. We therefore take Eq. (7.2) as reference for an order-of-magnitude estimation of the precession frequency and keep in mind that other modes exist.

7.2.2 Hysteresis

When we ramp down the external field, the spins still stay aligned with their neighbors, but due to dipolar interactions, a zero net magnetization would be energetically favorable. For this reason, domains form in the ferromagnet, which have different magnetic orientations and therefore reduce the overall magnetization, while still preserving a parallel spin orientation within the domain. In reality, the domain wall motion and reordering is dominated by material defects, leading to a magnetization depending on the magnetic history of the sample and therefore a hysteresis [61]. A sketch of the magnetization with respect to the applied magnetic field is given in Fig. 7.2. Three quantities are here of particular interest: The saturation magnetization M_{S} which is observed when all domains are aligned parallel to the external field, the remanence magnetization M_{R} which remains when setting the external field to zero, and the coercive field H_{C}, which has to be applied to reach zero magnetization again.

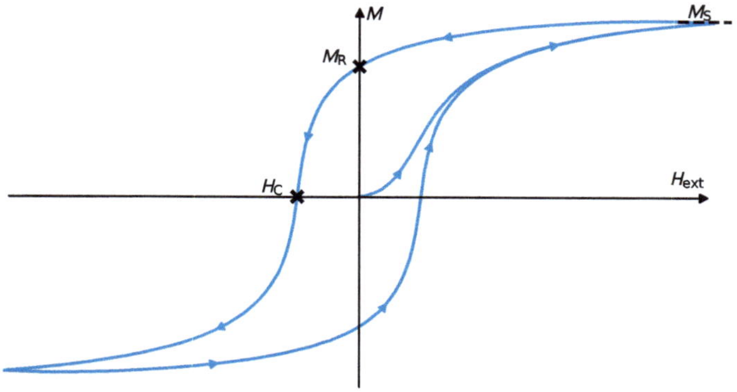

Figure 7.2: Magnetic hysteresis. The demagnetized sample starts at the origin and the magnetization increases with increasing applied field, until the saturation magnetization M_S is reached. When the magnetic field is decreased, the magnetization shows a hysteresis and only slowly decreases again. For $H_{ext} = 0$, a remanent magnetization of M_R remains, and the magnetization only reaches zero, when the coercive field $H_{ext} = H_C$ is applied. The shape of this curve was calculated by the Jiles-Atherton formula [182] with symbolic values and does not represent the measured samples.

For the experiments presented in Sec. 7.5, we keep in mind that the magnetization changes quickly close to the coercive field when the field direction is changed and therefore the oscillation frequency is very sensitive to the magnetic field. Due to the formation and switching of different domains in this area, we do not expect a well defined mode spectrum. It is therefore beneficial to change the magnetization as little as possible by magnetizing the sample with a strong magnetic field and performing only low-range sweeps afterwards without changing the sign of the external field.

7.3 Hybrid systems

If we want to couple a magnonic system with a cavity or a qubit, we need to align the fields such that the oscillating AC field of cavity or qubit is perpendicular to the static magnetic bias field. Then, the AC field can excite the precessional motion of the spins, leading to a coupling of the two systems. This can be realized in 3D cavities or planar geometries, which is detailed in the following

7.3.1 Cavity magnonics

Using a macroscopic microwave cavity for the coupling with magnetic oscillations is advantageous for the first experiments: The spatial extent of a 3D microwave cavity assures a homogeneous AC field distribution over the magnetic sample in terms of field strength and field direction. This eases theoretical considerations as well as numerical calculations, and allows to give a clear number of participating spins. The magnetic sample is typically of spherical shape, as this reduces the demagnetization field and creates fewer spatial magnon modes than rectangular shapes. These effects are beneficial when exciting the $\vec{k} = 0$ mode, the so called Kittel mode, and therefore help to create a simple and predictable mode spectrum. Also, the modular system, where cavity and magnetic medium can be easily exchanged, enables to study the exact same samples under different configurations.

Most research in the field of cavity magnonics has hence been performed with magnetic spheres placed in the anti-node of the cavity's AC magnetic field [62, 64, 183, 184]. These experiments opened new insights, but also gave rise to new research fields and raised new questions in the field of cavity magnonics.

7.3.2 Cavity quantum magnonics

To study these systems in more detail at low excitation powers, a natural approach is to introduce a nonlinear element, which allows to resolve the number of quantum excitations. This is realized by a superconducting qubit, placed in an antinode of the cavity's electric field. Via a common coupling to a cavity mode, a virtual qubit-magnon coupling is established [67], and single magnon resolution was demonstrated [68]. This magnon counting can be equally regarded as the dispersive shift of the qubit frequency by the population of the magnonic system or by an AC Stark shift of the qubit due to the AC signal of the magnonic excitation. These studies render the first quantum resolved measurements on a magnonic system. In follow-up experiments, also a time resolved control and readout of the magnon-qubit interaction was demonstrated [69, 140].

Especially with regard to the inclusion of a superconducting qubit, the 3D approach simplifies several experimental issues: First, the range in applicable magnetic fields is increased, since copper cavities can be used. Fabricated with a sufficiently high quality factor, these cavities do not alter in a magnetic field, in contrast to a superconducting resonator, as found in Sec. 6.3. The increased field range also expands the range of applicable magnetic materials, as discussed in 7.4.1. Second, due to the spatial distance between qubit and magnetic material, the bias magnetic field can be generated locally at the magnetic sample. The qubit can be additionally shielded from the magnetic bias field by permalloy and superconducting shields. And third, the modular approach again enables to characterize the qubit-cavity and magnon-cavity system independently before combining them. In addition, individual components can be exchanged, if their frequencies have to be adjusted, while the other components stay unmodified.

Although this 3D approach is useful for first experiments, it has several drawbacks: Especially with respect to scaling and future applications, cavities have a large spatial extent and cannot be easily integrated. In addition, due to the narrow AC field distribution of the qubit, only a virtual, cavity mediated coupling between qubit and magnon can be achieved and a direct coupling is not possible in a 3D geometry.

7.4 Direct qubit magnon coupling

These drawbacks can be solved by a 2D approach, where a planar qubit-resonator system is coupled to a magnetic system. In contrast to the 3D approach, multiple systems can be placed on one chip with a reduced cross-talk and the overall dimensions for one qubit-magnon-cell are drastically decreased. In planar geometry, it is also more straight-forward to create a tunable qubit, which can be tuned by fast flux pulses and thus allows for a dynamical coupling. With the possibility to apply $\hat{\sigma}_z$ pulses to the qubit, direct entanglement between qubit and magnon states can be reached and the coupled system can be brought into a superposition state. The planar approach also opens the possibility to directly couple the qubit to the magnetic material and avoid the detour via coupling cavity modes.

For a direct coupling, the magnetic material has to be placed in the maximum of the qubit's AC magnetic field. This corresponds to the areas with the highest current density in the qubit and hence the leads to the junction. The alternating current across the JJ creates an oscillating magnetic field, which can in turn excite the precession of the spins in a magnetic system on top of the qubit.

7.4.1 Choice of magnetic material

One of the most frequently used magnetic materials for coupling experiments in the microwave regime is YIG[1], which is a ferrimagnetic insulator. Its very low losses and the therefore small linewidth are beneficial for resonance experiments, making YIG a very promising material for coherent information processing. For 2D experiments, it was recently investigated in pre-characterization experiments with superconducting resonators [65, 185].

A disadvantage for YIG is the need of GGG[2] as a substrate for the crystal growth process. This substrate shows strongly increased losses at low temperatures [186], which

[1] YIG: Yttrium Iron Garnet
[2] GGG: Gadolinium Gallium Garnet

counteracts the low-loss properties of YIG. To circumvent this problem, YIG spheres are used in 3D applications, which are fabricated from bulk YIG. Alternatively, the substrate can be polished away after fabrication, which is technologically challenging and influences the mechanical stability of the sample.

One additional and much stronger constraint is given by the field-dependent resonance frequency from Eq. (7.2). If we take literature values of $\gamma = 28\,\mathrm{GHz/T}$ and $\mu_0 M_\mathrm{S} = 175\,\mathrm{mT}$ [187, 188], we see that we need external magnetic fields on the order of $B_\mathrm{ext} \geq 100\,\mathrm{mT}$ to create resonance frequencies at $\omega_\mathrm{P} \approx 5\,\mathrm{GHz}$. This is clearly incompatible with the maximum magnetic field for a superconducting qubit, studied in Ch. 6.

To decrease the bias magnetic field while still reaching resonance frequencies in the gigahertz range, we can increase the saturation magnetization according to Eq. (7.2) by choosing a different magnetic material. The increased saturation magnetization is typically accompanied with an increased spin density ρ_S, which increases in turn the coupling strength after Eq. (7.5). Permalloy, an iron-nickel alloy, is reported to have a high saturation magnetization and was recently investigated in combination with superconducting resonators [189, 190]. From the Slater-Pauling curve, which plots the atomic saturation magnetization versus the number of electrons per atom [191], we find an even higher maximum between the ferromagnets iron and cobalt. More detailed studies of $\mathrm{Fe}_x\mathrm{Co}_{1-x}$ alloys find a maximum of the saturation magnetization for $x \approx 30\,\%$ [182].

In pre-characterization measurements [188], we measured a sample fabricated by an external group, where unfortunately the stoichiometry and fabrication methods are widely unknown. For this sample, we found a saturation magnetization of $\mu_0 M_\mathrm{S} = 2.1\,\mathrm{T}$ for FeCo. Together with a gyromagnetic ratio of $\gamma = 28.2\,\mathrm{GHz/T}$ we find that a magnetic field of $B_\mathrm{ext} = 15\,\mathrm{mT}$ suffices to bring the resonance frequencies to $\omega_\mathrm{P} = 5\,\mathrm{GHz}$. The measured data are shown in Fig. 7.3 together with a fit to Eq. (7.2).

From these data, we can also extract the linewidth of the magnetic resonance, which corresponds to the damping of the spin wave oscillation. In the relevant frequency range of 4-8 GHz, we find on average $\kappa_\mathrm{m}/2\pi = (350 \pm 19)\,\mathrm{MHz}$ at room temperature. In the literature however, $\mathrm{Fe}_x\mathrm{Co}_{1-x}$ films with an ultra-low magnetic linewidth are reported for $x \approx 25\,\%$, resulting in $\kappa_\mathrm{m}/2\pi = 56\,\mathrm{MHz}$ at $\omega_\mathrm{m}/2\pi = 10\,\mathrm{GHz}$ [192].

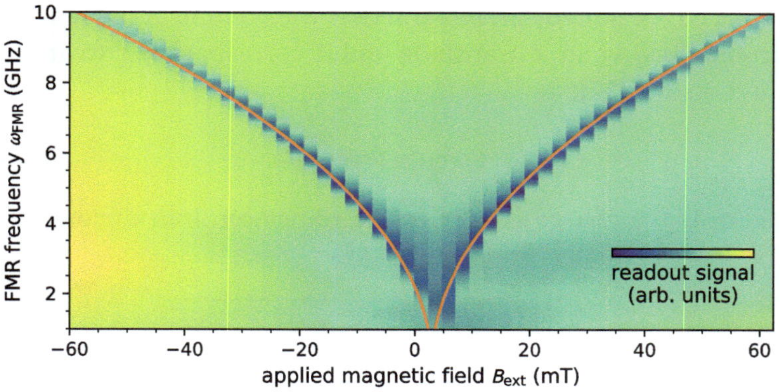

Figure 7.3: FMR measurements on an externally fabricated sample. Results from a pre-characterization study on a FeCo sample. The orange line shows the theoretical curve for $\gamma = 28.2\,\mathrm{GHz/T}$ and $\mu_0 M_S = 2.1\,\mathrm{T}$ together with an offset field of $B_{offs} = 3\,\mathrm{mT}$. The sample used for this measurement is a thin FeCo film, fabricated in an external lab with unknown stoichiometry and fabrication procedure, and not lithographically structured. Measurement data and fit values are taken from Ref. [188].

7.4.2 Coupling strength calculation

To calculate the coupling of a spin ensemble to the qubit, we start with the individual coupling strength g_i of each spin. By applying perturbation theory to the Tavis-Cummings Hamiltonian [193], we obtain for the total coupling of the system

$$g_\Sigma = \sqrt{\sum_i^N g_i^2}. \tag{7.3}$$

For the case of equal coupling strength, $g_i = g$, this results in the well-known relation $g_\Sigma = g\sqrt{N}$. In our case however, we assume that all spins S are equal, but feel a different local magnetic field $B_1(\vec{r}_i)$ at their location $\vec{r}_i$. The index 1 emphasizes that the field is created by one excitation in the qubit. Under these circumstances, it holds $g_i = \gamma S B_1(\vec{r}_i)$, where γ is the gyromagnetic ratio.

We can calculate the field distribution $B_1(\vec{r}_i)$ with numerical simulations from the known sample geometry. Due to the limited precision of numeric simulations, it can be disadvantageous to simulate $B_1(\vec{r}_i)$ at the qubit current I_q, which corresponds to a single excitation in the qubit and is on the order of several nanoampere. Instead, we assume $B \propto I$, since the AC field can be considered as small perturbation to the DC bias field. We therefore use a higher current I_{sim} for the simulation, which also makes the numeric simulation independent from the qubit parameters. We therefore obtain:

$$g_\Sigma = \gamma S \frac{I_q}{I_{sim}} \sqrt{\sum_i^N |B_1(\vec{r}_i)|^2} = \gamma S \frac{I_q}{I_{sim}} \sqrt{\frac{N}{V_S} \int |B_1(\vec{r}_i)|^2 \,\mathrm{d}V_S}, \tag{7.4}$$

where we replaced the sum by an integral over the volume of the magnetic sample, V_S. The current flowing in a transmon qubit corresponding to its first excited state is found to be [17]:

$$I_q = \omega_{01} e \sqrt{\hbar\omega_{01}/2E_C},$$

where E_C again is the charge energy of the transmon. Introducing the spin density $\rho_S = N/V_S$, we finally find

$$g_\Sigma = \gamma S \frac{\omega_{01} e}{I_{sim}} \sqrt{\frac{\hbar\omega_{01}}{2E_C} \rho_S \int |B_1(\vec{r}_i)|^2 \, dV_S}. \tag{7.5}$$

When comparing Eq. (7.5) to analytic expressions for the coupling strength found in the literature [62, 194], we see that these formulas involve quantities such as the mode volume $V_a = \int |B|^2 dV / |B_{max}|^2$ or a spatial overlap η instead. These quantities are useful for the 3D case, where the cavity volume is confined and the magnetic field can be regarded as being constant over the magnetic material. For our 2D case, the integration boundaries for $\int |B|^2 dV$ are in principal infinity (in contrast to the confined space of dV_S), which makes a numeric integration slow, and the field distribution at the magnetic material would not be respected.

7.4.3 Simulation of the coupling strength

The magnetic field distribution is simulated by 2D finite element simulation software, where we used FEMM[3] for our purposes. For the simulations, we varied the spacing between the two chips, given by the resist pad thickness, and the width and thickness of the magnetic film. The results are shown in Fig. 7.4 a) and b). We find that increasing the width of the magnetic structure further than $w \approx 15\,\mu m$ does not significantly increase the coupling strength, as long as the chip spacing is small. In general, we try to reduce the size of the magnetic sample, as a dissipative metal close to the qubit gives rise to microwave losses. For the experiment however, a structure with $w = 40\,\mu m$ is used, which is easier to align onto the qubit junction area. Together with a chip spacing of $d = 1\,\mu m$ and a film thickness of $h = 400\,nm$, we calculate a coupling strength of $g = 171.2\,MHz$.

[3] Finite Element Method Magnetics, version 4.2, femm.info

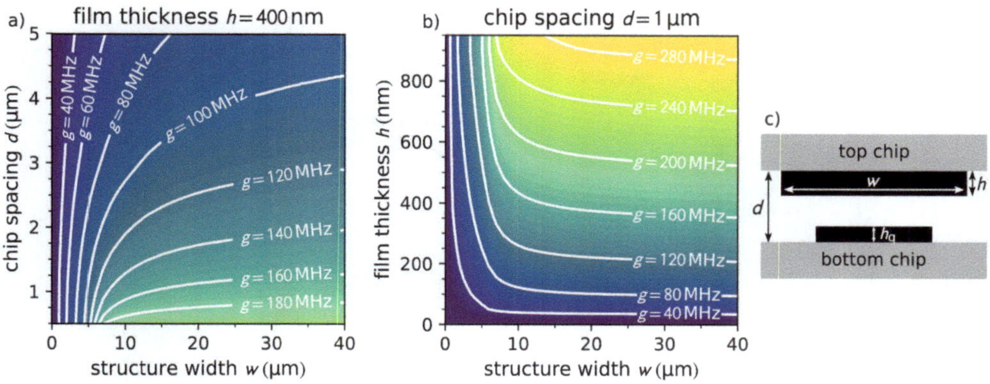

Figure 7.4: Simulated coupling between qubit and magnetic material. The coupling strength is calculated a) for different chip spacings d and b) for different film thicknesses h. A schematics of the simulated setup is given in c). The magnetic field distribution of the qubit is calculated and used for calculating the coupling strength after Eq. (7.5). As expected, the coupling increases with increasing width and thickness and therefore increasing number of spins, and decreases with increasing distance between the chips. For $w \gtrsim 15\,\mu\text{m}$, the coupling only slightly increases for a narrow chip spacing. Note that the chip spacing is measured between the substrates, i.e., for $d = h + h_q$, the films would touch, where $h_q = 100\,\text{nm}$ is the thickness of the qubit metallization.

7.4.4 Flip-chip approach

As seen from the simulations, we want to reduce the spacing between qubit and magnetic material to increase the coupling strength. A fabrication of both structures directly on top of each other would in principal be possible, but is accompanied by several technical challenges, i.e., damage of the Josephson junction in the deposition process of the magnetic material or contamination of the superconductor fabrication facilities with magnetic residues. Therefore, we chose to realize the coupling of the two systems with a flip-chip approach. Here, the two structures can be fabricated independently on individual chips and then placed face to face in a sample box. This design was already used for other coupling experiments with transmon qubits [102]. The sample box used there includes two rectangular chips which are rotated by 90° against each other. The distance between the chips is fixed by posts inside the sample box, which have a pre-defined step height and result in a minimum chip spacing of $d = 35\,\mu\text{m}$, due to the limited resolution of the CNC mill in our workshop. As seen in Fig. 7.4, this distance is still too big for our application.

The second approach was therefore to reduce the size of the top chip and take microstructured pads of optical resist as spacers between the chips. This avoids electrical contact as well as the creation of scratches in the structures in the alignment process of the chips. Since transparent sapphire is used as a substrate for the

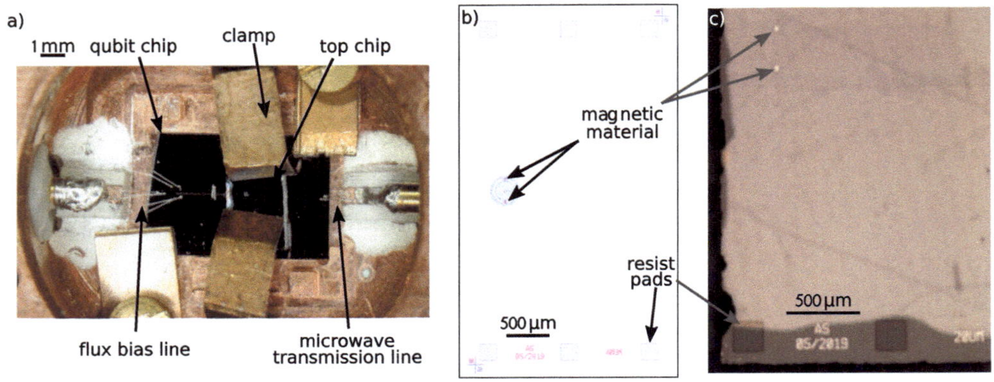

Figure 7.5: Flip-chip approach. a) Photograph of the sample box with the qubit chip (black) in the center and the top chip (transparent) with the magnetic material. The box features two microwave connections, used for microwave readout and flux tuning of the qubit. The two chips are held in place by copper-beryllium clamps, which stay flexible at low temperatures. Design b) and micrograph c) of the top chip. A zoom-in of the design view is given in Fig. 7.6. The magnetic material is not centered on the chip area but shifted to the left, such that the top chip covers all of the underlying readout resonator (not shown). Otherwise, the resonator's impedance would abruptly change in its center and additional modes would occur. In the dicing process, the substrate breaks off at the edges, which can be seen in c), but does not influence the functionality of the chip.

upper chip, the magnetic structure can be aligned under an optical microscope to be above the junction area. The resist pads are structured by an optical mask and the resist is baked on a hotplate after development, making it durable against organic solvents. From typical fabrication parameters, we get a resist film thickness of $d \approx 1.2\,\mu\text{m}$. The pads are positioned to be as far away as possible from both qubit and resonator to reduce additional microwave losses caused by the resist. The top chip is therefore only supported at its edges (see Fig. 7.5), leading to a bending in its center. This is increased by the clamps, which hold the top chip in place when mounting the sample in the solenoid.

A manual adjustment of the chip positions with an accuracy in the range of few micrometers with the help of tweezers and a microscope is possible, as shown in Fig. 7.6 a). Although sapphire is a hard material, the chips touch already at a bending of only 0.05 % for a chip length of 4.4 mm, which can lead to scratches on all structures and eventually a galvanic contact. Great diligence is therefore needed in the alignment process to avoid damage as depicted in Fig. 7.6 c), where clear scratches across the qubit islands and the junction leads are visible.

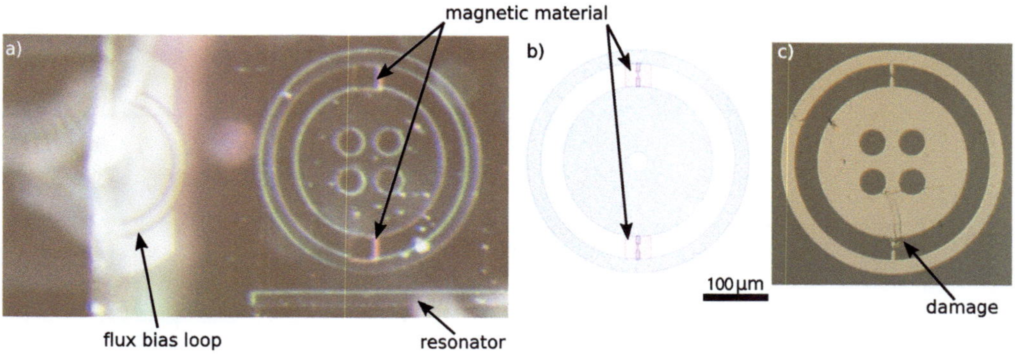

Figure 7.6: Alignment of the top chip. a) Micrograph of the chip stack. Since the top chip is transparent, the alignment of the magnetic material on the qubit leads can be adjusted under a microscope. b) Design schematics of the stacked system. The qubit islands are depicted in blue, the junction leads in pink, and the magnetic material in red. c) Great care has to be taken when aligning the chips, since the structures can be damaged if too much force is applied. A deep scratch in the center island and the lower junction lead is visible after an unsuccessful alignment approach. The scale bar is valid for all subfigures.

7.5 Measurements

We installed the flip-chip sample in the cryostat and measured its properties at millikelvin temperatures under the influence of a static magnetic field. In the following, the measured data on the resonator and qubit frequencies are presented, followed by a discussion about the origins of the observed spectra.

When we apply a magnetic field to tune the resonance frequency of the magnetic material, we have to keep in mind that FeCo shows a hysteretic behavior. We therefore expect more reproducible and reliable results when we do not sweep through the whole hysteresis loop, since domain wall shifting and domain switching could still occur in the frequency range of interest. Instead, we do not change the sign of the magnetic field for the following measurements and try to keep the change in magnetization as small as possible. If not noted differently, for the measurements presented in the following, we ramp the magnetic field to a value of $B = 81\,\mathrm{mT}$ for a short time, ramp the field down to 0 and cycle the base temperature above $T = 5\,\mathrm{K}$ to release all trapped flux in the superconductor.

7.5.1 Resonator measurements

When ramping up the magnetic field, we see several avoided level crossings in the resonator frequency. They show different coupling strengths, different slopes of the magnetic system's frequency with respect to the magnetic field, and even

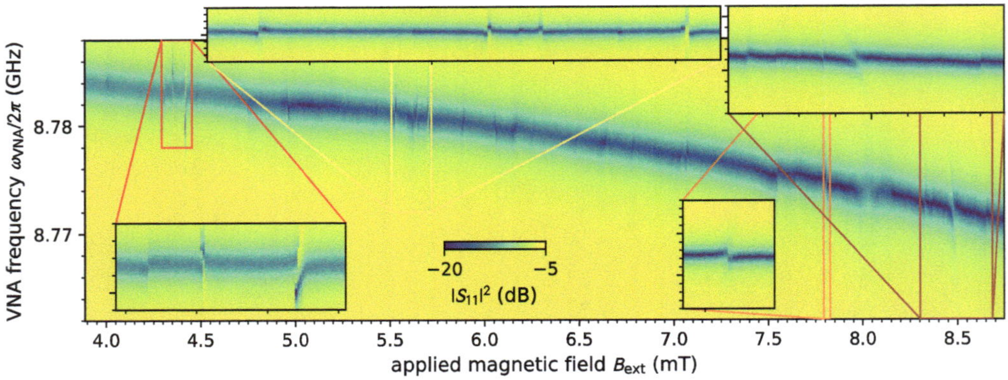

Figure 7.7: Avoided level crossings in the resonator spectrum. The plot shows color-coded the amplitude of the reflected signal to find the resonance frequency of the resonator for different applied magnetic fields. The data are taken on an upsweep of the magnetic field. The insets show zoom-ins on different features we observed in the resonance spectrum. Unfortunately, the measured signal is too weak to extract a coupling strength for the different features by means of a fit, but we can estimate the coupling strength to be on the order of a few megahertz.

different signs of the slope. One measurement is shown in Fig. 7.7 and additional data can be found in Appendix C. The coupling strength is just large enough to see the gap in the spectrum and the onsets of the anticrossing, but not large enough to track the resonance frequencies of the hybridized system. Therefore, and due to sudden jumps in the resonance spectrum, we are not able to fit the anticrossing and extract a coupling strength. Tracing the lines by eye, we can estimate the coupling strengths to be in the range of a few megahertz.

Although the resonator is not directly coupled to the magnetic material, we would theoretically expect a weak coupling between both systems. This is on the one hand due to the large field distribution of microstrip resonators and on the other hand through the qubit-mediated coupling. The latter can be calculated by $g_{RM,eff} = (g_{QR}g_{QM})/(\omega_R - \omega_{01})$ [195], where g_{QR} (g_{QM}) is the coupling strength between qubit and resonator (magnetic material), and is expected to be in the range of a few megahertz.

These data show a variety of magnetically tunable resonances, with coupling strengths in the same order of magnitude as expected from a spurious coupling between magnon and resonator. The variety of different modes would be explainable by taking into account the mode spectrum of the magnetic sample, as discussed in Sec. 7.5.3.

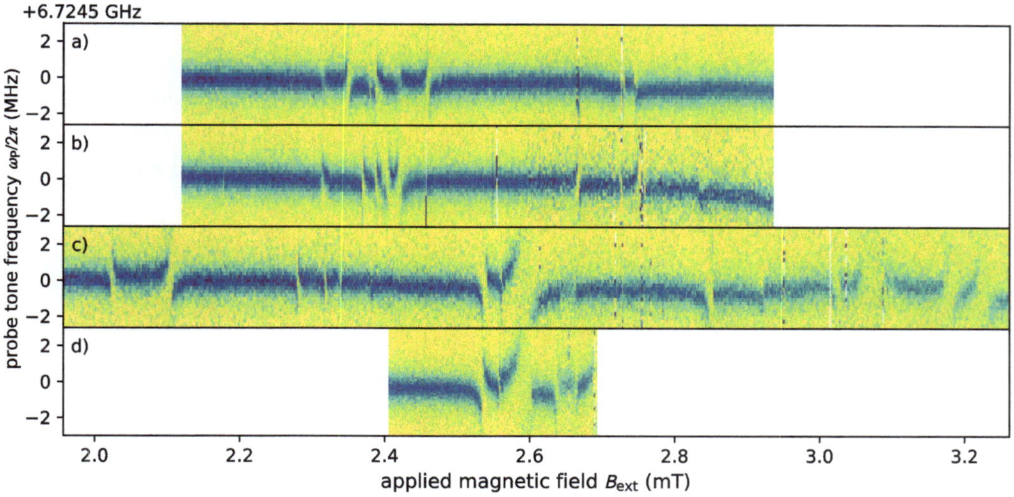

Figure 7.8: Four qubit sweeps in the magnetic field. Color-coded is the normalized dispersive readout signal in arbitrary units. White points represent masked data, where the resonator signal was not reliable. The order of the subplots corresponds to the order of measurement. It can be clearly seen that several features re-appear at comparable fields (2.3 – 2.4 mT), but are different on each sweep. Other features, like the anticrossing at 2.58 mT, appear only in sweeps c) and d).

7.5.2 Qubit measurements

Scanning the qubit frequency, we find several avoided level crossings in its spectrum already at very low magnetic fields. Again, they differ strongly in coupling strength, slope and the sign of the slope. In total, they do not correspond to the magnetic resonance we expected. Due to the slow drifts of the individual resonances as well as sudden jumps, it is not possible to predict the exact position of an anticrossing for a new measurement. Together with a long data acquisition time, we are hence not able to resolve the anticrossings with a higher precision.

The data of several qubit scans are shown in Fig. 7.8. The described sequence of ramping the solenoid field to $B_{\text{ext}} = 81\,\text{mT}$ and afterwards cycling the fridge was executed for subfigures c) and d).

A fit of the two unperturbed anticrossings in Fig. 7.8 c) and d) is given in Appendix C. This is however subject to large errorbars, as we cannot see both branches of the avoided level crossing simultaneously due to the weak coupling strength. From the visible data we can make a rough estimate for the prominent avoided level crossing at $B_{\text{ext}} = 2.59\,\text{mT}$ and find a lower bound of $g/2\pi \gtrsim 4\,\text{MHz}$.

This order of magnitude estimation can be enhanced in future measurement, which include a drive tone close to the qubit frequency. Now that the position of the avoided level crossings is approximately known, the drive tone frequency can be scanned close to the qubit frequency. As described in Ch. 5 we then would expect a stronger AC Stark shift of the qubit levels when the drive matches the resonance frequency of the magnetic system, due to the resonant amplitude enhancement of the microwave signal. This would enable to trace the resonance lines for larger detunings and subsequently narrow down the estimate of the coupling strength.

7.5.3 Discussion

The acquired data on resonator and qubit frequency show a clear interaction between resonator, qubit, and multiple modes of a separate system, whose resonance frequencies can be tuned by a magnetic field.

From theoretical and measured properties of the FeCo film, we would expect the magnetic resonance to reach the qubit frequency only at stronger magnetic fields on the order of $B_{\mathrm{ext}} \approx 20\,\mathrm{mT}$. Possible explanations for this discrepancy are the existence of additional shape and magnetocrystalline anisotropy fields, which we ignored in the theoretical considerations in Sec. 7.2.1. In addition, the hysteresis curve and especially the remanent magnetization are not known for the magnetic films used for the experiment.

Depending on the orientation of the external field to the film, the shape anisotropy can also counteract the applied magnetic field [181], which can result in a decreasing resonance frequency with increasing field. This behavior can be used to explain the anticrossings for which the second system's frequency goes down with increasing magnetic field.

In our theoretical considerations, we only took into account the uniform mode with $\vec{k} = 0$. Due to the inhomogeneous field created by the qubit leads, it would also be possible to excite magnon modes with $\vec{k} \neq 0$, which in general show resonance frequencies depending on their wave vector, $\omega_{\mathrm{m}}(\vec{k})$. This may be an explanation for the great variety of different modes we observe in our spectrum, each with their own coupling strength.

Efforts to further verify these explanations, such as simple FMR measurements and a SQUID magnetometer did not give new insights, since the sample size and the number of spins was too small for the sensitivity of the measurement devices. These measurements could therefore neither confirm nor refute the theory that the avoided level crossings are hybrid resonances with the magnetic system.

From Sec. 3.6 we know that an avoided level crossing is only visible when all included linewidths are similar to or smaller than the coupling strength. In the measured data however we see the onsets of avoided level crossings, but cannot trace the individual branches far enough to see both branches simultaneously. We therefore can follow that the largest linewidth is on the same order of magnitude as the coupling strength. With a lower boundary for $g/2\pi = 4\,\mathrm{MHz}$, as estimated from the qubit spectrum, together with the FMR linedwith previously measured on a different sample, $\kappa_\mathrm{m}/2\pi = (350 \pm 19)\,\mathrm{MHz}$ [188], we clearly would have to follow that the measured anticrossings cannot result from magnon oscillations. If we however take the value of $g/2\pi \sim 110\,\mathrm{MHz}$ which we get from the avoided crossing fit in Appendix C and assume a high-quality FeCo film with $\kappa_\mathrm{m}/2\pi < 56\,\mathrm{MHz}$ [192], the anticrossing would be well resolvable.

Since the data basis is rather weak and the observed crossings do not match our intuitive expectation of an anticrossing with one single straight line, as expected from the FMR signal shown in Fig. 7.3, we want to emphasize that this chapter makes no claim about the origin of the observed anticrossings. Instead, we can only state that we see a variety of resonances coupling to our qubit, which can be tuned by a magnetic field. It is well possible that these resonances do not have their origin in the magnetic material we placed on top of the qubit, but result from other effects, like a changing flux vortex configuration or, e.g., magnetically tunable defects. We want to note the following observations from our experiments and leave the conclusion up to the interested reader:

- We could not observe anticrossings when the magnetic field was swept in a bipolar sweep (i.e., from positive to negative values or vice versa). The crossings only reappeared after applying a strong magnetic field in one direction. This supports the idea of having a hysteretic system.
- This hysteresis "survived" a thermal cycle of the sample above the superconducting transition temperature, and we found anticrossings at comparable field strengths repeatedly. This negates the idea that the resonances come from an interaction of the superconductor itself with the magnetic field.
- In the pre-characterization of the qubit chip without a magnetic material on top, we did not see any anticrossing. There however, we did not perform the sequence of applying a strong magnetic field, cycling the base temperature and scanning the magnetic field with a fine resolution. The absence of avoided level crossings without the magnetic chip on top support the theory that the resonances arise from magnon oscillations.

7.6 Conclusion and outlook

In this section, we presented the basic mechanism to achieve the coupling between a superconducting qubit and a magnetic material. We derived an analytical formula to calculate the coupling strength for a given geometry from the results of finite elements simulations. In the last part, we presented measured data on the coupled system, which show a rich set of features with different magnetically tunable resonances. The origin of these resonances could not be traced back unambiguously to the magnetic alloy on top of the qubit, but opens a new door to further analyze these resonantly coupled features.

Based on the knowledge acquired from the measurements in this chapter, several simple experiments can be performed that could help to find the origin of the magnetically tunable resonances. These experiments include the measurement of a bare qubit in a magnetic field following the described measurement scheme, additional FMR measurements of the magnetic sample at low temperatures and the fabrication of new magnetic samples with a larger size, which could be characterized by other methods.

If these additional measurements confirm the appearance of multiple magnon modes at the qubit frequency in an external field of a few millitesla, and also demonstrate the low linewidth of the selected FeCo alloy at low temperatures, the measurements presented in this chapter would render one of the first experimental demonstrations of a direct qubit-magnon coupling. On this system, more measurements like coherence analysis and noise spectroscopy could be performed to find out more about the magnon system. Replacing the qubit chip by a tunable version, together with a stabilized magnetic bias field, would enable to study the magnon spectrum on a larger frequency scale and find more information about the frequency-dependent coherence of the magnonic system. Eventually, a fast tunable qubit would enable to create entanglement between qubit and magnon states and bring the system in a superposition.

For the case that the additional measurements demonstrate that the magnetically tunable resonances are independent of the magnetic sample on the qubit, the presented data would open the door to a new field of research. One likely option in this case would be that defects on the qubit, which have been studied in great detail for electric fields and strain, are also sensitive to magnetic fields. This effect can then be used to study these defects in more detail and learn more about one of the most prominent loss channels in superconducting qubits.

8 Conclusion

The goal of this thesis was to utilize a superconducting quantum bit for quantum sensing experiments and pave the way for the investigation of magnetic excitations in the quantum regime. By developing a new sensing scheme for oscillating microwave signals, studying a superconducting qubit in a magnetic field and interfacing the qubit with a ferromagnetic material, we created a solid foundation for future experiments, which allow to study magnons with a quantum sensor.

For our implementation of a quantum sensor, I started this thesis with the development of a new sensing scheme, based on the AC Stark effect of an anharmonic multilevel quantum system [71]. By comparing the shift of higher excited qudit states with a model system, this scheme allows to characterize a microwave signal simultaneously in amplitude and frequency. Utilizing a fixed-frequency superconducting qubit, we demonstrated the sensitivity in a remarkable frequency range of more than one gigahertz and for a great range in signal strength. Outstanding in this sensing scheme is the locality of the measurement, which allows to measure signals arriving at the qudit position *in situ*. This is of unique advantage for the development, characterization and optimization of quantum circuits, as it enables to measure crosstalk and spurious couplings between different circuit elements. As demonstrated, our scheme allows to characterize the frequency-dependent microwave transmission function from the room temperature measurement equipment to the qubit at low temperatures. This calibration can be executed under the exact same ambient conditions as the actual experiments, giving valid calibration data. This includes temperature-dependent cable resonances as well as the frequency-dependent coupling between qubit and microwave transmission line, which is otherwise inaccessible. The presented sensor is also helpful for the development of hybrid quantum systems, where it allows for the characterization of excited resonance modes coupled to the qudit.

The measurements with this sensor have been carried on in our group with a focus on increasing the sensor precision [72]. Utilizing time-resolved data acquisition, the change in qudit frequency was determined with higher accuracy, leading to an improved resolution of the sensor in both amplitude and frequency.

Continuing the way towards enabling quantum sensing of magnetic excitations, we fathomed the sensor's limits in compatibility with a magnetic field. To this end, we studied both frequency and coherence properties of a superconducting transmon qubit in an in-plane magnetic field. Although superconductivity suffers under this influence, we were able to demonstrate a quantum coherent behavior up to remarkably high magnetic field values of $B \approx 40\,\mathrm{mT}$, which was beyond the expected regime and is four times larger than the critical field of bulk aluminum [70]. These findings build the foundation for quantum sensing experiments with superconducting qubits in magnetic fields and shine a promising light on future applications.

Analyzing our measured data, we found a second, spurious Josephson junction in the qubit to be relevant. This stray junction exists due to the fabrication process and is commonly neglected. Having a large cross section in the magnetic field, this additional junction gives rise to magnetically induced fluctuations of the qubit transition frequency. Although our calculations and measurements show that this spurious junction does not limit the dephasing time of our sample in the case of small applied fields, the situation changes for samples with larger junction cross sections and larger coherence times. Meanwhile, these spurious or parasitic junctions have gained additional research interest in independent studies, where they were found to host atomic defects, which increases decoherence independent of the magnetic field [37].

In the last experimental chapter, we combined the previous findings to build a hybrid chip, consisting of a superconducting quantum bit and a ferromagnetic structure. By analytical considerations and numeric simulations of the geometry, we estimated the coupling strength and performed pre-characterization measurements of similar magnetic materials. In the final experiment, where we directly interfaced qubit and magnetic material at millikelvin temperatures, we observed a rich set of interesting features when tuning the magnetic field. Several onsets of avoided level crossings between qubit and magnetically tunable resonances were visible, suggesting that only minor adjustments are needed to reach the strong coupling regime. The origin of this great variety in different magnetically tunable modes is still subject to investigation, but the presence of different magnetostatic modes is seen as a very plausible explanation. These promising data inspire the further investigation of qubit-magnon hybrid quantum systems and demonstrate the feasibility of a direct coupling approach in planar geometry.

Outlook

Quantum sensing with superconducting qubits has many near- and mid-term applications, especially with respect to improving quantum computation devices and developing new hybrid quantum systems. Already now, the sensing scheme presented in this thesis can be used to quantify microwave crosstalk in quantum chips with multiple qubits, resonators and microwave feed lines. Based on the AC Stark shift, this sensing scheme can be applied without the need for pulsed microwave measurements or tunable qubit samples. Due to this astonishing simplicity, the transmon sensor is an ideal device to enable, e.g., quality assurance in the rapidly growing business field around cryogenic quantum technologies. Compared to a standard cryogenic characterization of microwave components using a vector network analyzer, only a transmon and two low-quality microwave sources are needed, without any additional cryogenic wiring. This would enable the growing number of suppliers for high-quality cryogenic microwave components, i.e., microwave cables, attenuators, circulators, and amplifiers, to provide a full cryogenic calibration at acceptable cost, which renders a unique selling point.

Based on the broad field of current applications for SQUID sensors, which are used already now for the sensing of magnetic fields created by the human body in biomedical applications or the detection of cosmic radiation by magnetothermal effects, many sensing experiments are conceivable. The scope of these applications can be enhanced by replacing the SQUID with a superconducting qubit, and eventually introduce the sensitivity of quantum coherence and entanglement. Using methods like dynamical decoupling and spatial entanglement, the sensitivity to temporal and spatial fluctuations can be dynamically adjusted, opening new prospects on measurement accuracy and flexibility.

The promising results we found in the experiments towards a coherent coupling of superconducting qubits and magnons allow for a positive outlook on future hybrid quantum systems. Enabling a coherent energy transfer between qubit and magnon system, the quantum mechanical dissipation mechanisms in magnons could be studied in-depth by relaxometry studies and phase coherence measurements, to find reasonable countermeasures. The entanglement of qubit and magnon states, and qubit-assisted relaxation time measurements for single magnons would pave the way for further investigations of magnonic systems with respect to applications in quantum storage devices or the realization of a quantum bus. In combination with the large variety of classical magnonic computation building blocks and systems like magnonic metamaterials, a versatile toolbox can be created to advance the implementation of quantum effects in computation.

The great scientific interest in enabling a quantum-resolved study of magnon excitations can be seen by the number of experiments published during the course of this thesis. Studying magnons in a 3D cavity at low temperatures, we [62, 64] and other groups [183] investigated the temperature and power dependence of internal losses in YIG, a commonly used ferrimagnet. A superconducting qubit was soon added to this system [66], but only very recently, a coherent control of the magnon states was achieved in this cavity-mediated 3D approach [69]. Following a planar realization, combinations of planar superconducting resonators with ferromagnets have been studied at the same time [185, 194], and recently also other magnetic materials, which are more promising for a planar combination with a superconducting qubit, reached the focus of several research groups [189, 190].

Simultaneously, the magnetic field compatibility of superconducting qubits has gained more research attraction. Here, new materials are investigated, such as semiconducting nanowires [167] and disordered superconductors for the junction [196] as well as for the whole qubit [197]. Both approaches promise to increase the range in magnetic fields, in which superconducting qubits can be used and therefore lower the constraints for a direct coupling between qubit and different magnetic materials.

Based on these developments and our promising findings, we are convinced that a coherent coupling between superconducting qubits and ferromagnetic excitations is possible in a planar geometry and will be realized in the near future. Finally, the results of these quantum sensing experiments contribute to the development of quantum computers and enable new communication interfaces.

Bibliography

[1] F. R. Shapiro: *The Yale Book of Quotations*. Yale University Press, 2006. 1216 pp. URL: https://www.ebook.de/de/product/5481952/fred_r_shapiro_the_yale_book_of_quotations.html (cit. on p. 1).

[2] G. E. Moore: *Cramming more components onto integrated circuits*. Electronics **38**.8 (1965). URL: https://www.supplychain247.com/images/pdfs/gordon_moore_1965_article.pdf (cit. on p. 1).

[3] R. Schaller: *Moore's law: past, present and future*. IEEE Spectrum **34**.6 (1997), pp. 52–59. DOI: 10.1109/6.591665 (cit. on p. 1).

[4] W. M. Holt: *1.1 Moore's law: A path going forward. 2016 IEEE International Solid-State Circuits Conference (ISSCC)*. IEEE, 2016. DOI: 10.1109/isscc.2016.7417888 (cit. on p. 1).

[5] J. J. Liou, F. Schwierz, and H. Wong: *Nanometer CMOS*. Taylor & Francis Ltd., 2010 (cit. on p. 1).

[6] L. Eeckhout: *Is Moore's Law Slowing Down? What's Next?* IEEE Micro **37**.4 (2017), pp. 4–5. DOI: 10.1109/mm.2017.3211123 (cit. on p. 1).

[7] M. M. Waldrop: *The chips are down for Moore's law*. Nature **530**.7589 (2016), pp. 144–147. DOI: 10.1038/530144a (cit. on p. 1).

[8] N. Wirth: *A plea for lean software*. Computer **28**.2 (1995), pp. 64–68. DOI: 10.1109/2.348001 (cit. on p. 1).

[9] N. Lambert, Y.-N. Chen, Y.-C. Cheng, C.-M. Li, G.-Y. Chen, and F. Nori: *Quantum biology*. Nature Physics **9**.1 (2012), pp. 10–18. DOI: 10.1038/nphys2474 (cit. on p. 1).

[10] G. S. Engel, T. R. Calhoun, E. L. Read, T.-K. Ahn, T. Mancal, Y.-C. Cheng, R. E. Blankenship, and G. R. Fleming: *Evidence for wavelike energy transfer through quantum coherence in photosynthetic systems*. Nature **446**.7137 (2007), pp. 782–786. URL: http://dx.doi.org/10.1038/nature05678 (cit. on p. 1).

[11] R. P. Feynman: *Simulating physics with computers*. International Journal of Theoretical Physics **21**.6-7 (1982), pp. 467–488. DOI: 10.1007/bf02650179 (cit. on p. 1).

[12] L. K. Grover: *Quantum Mechanics Helps in Searching for a Needle in a Haystack.* Physical Review Letters **79**.2 (1997), pp. 325–328. DOI: 10.1103/physrevlett.79.325 (cit. on pp. 2, 12).

[13] P. W. Shor: *Algorithms for quantum computation: discrete logarithms and factoring. Proceedings 35th Annual Symposium on Foundations of Computer Science.* 1994, pp. 124–134. DOI: 10.1109/SFCS.1994.365700 (cit. on p. 2).

[14] F. Arute, K. Arya, R. Babbush, D. Bacon, J. C. Bardin, et al.: *Quantum supremacy using a programmable superconducting processor.* Nature **574**.7779 (2019), pp. 505–510. DOI: 10.1038/s41586-019-1666-5 (cit. on pp. 2, 13).

[15] E. Pednault, J. Gunnels, D. Maslov, and J. Gambetta: *On "Quantum Supremacy".* Tech. rep. IBM, 2019. URL: https://www.ibm.com/blogs/research/2019/10/on-quantum-supremacy/ (cit. on p. 2).

[16] Y. Nakamura, Y. A. Pashkin, and J. S. Tsai: *Coherent control of macroscopic quantum states in a single-Cooper-pair box.* Nature **398** (1999), pp. 786–788. DOI: 10.1038/19718 (cit. on pp. 2, 11).

[17] J. Koch, T. M. Yu, J. Gambetta, A. A. Houck, D. I. Schuster, J. Majer, A. Blais, M. H. Devoret, S. M. Girvin, and R. J. Schoelkopf: *Charge-insensitive qubit design derived from the Cooper pair box.* Physical Review A **76** (2007), p. 042319. DOI: 10.1103/PhysRevA.76.042319 (cit. on pp. 2, 20, 22, 23, 25, 70, 83, 110).

[18] V. E. Manucharyan, J. Koch, L. I. Glazman, and M. H. Devoret: *Fluxonium: Single Cooper-Pair Circuit Free of Charge Offsets.* Science **326**.5949 (2009), pp. 113–116. DOI: 10.1126/science.1175552 (cit. on p. 2).

[19] F. Yan, S. Gustavsson, A. Kamal, J. Birenbaum, A. P. Sears, D. Hover, T. J. Gudmundsen, D. Rosenberg, G. Samach, S. Weber, J. L. Yoder, T. P. Orlando, J. Clarke, A. J. Kerman, and W. D. Oliver: *The flux qubit revisited to enhance coherence and reproducibility.* Nature Communications **7**.1 (2016). DOI: 10.1038/ncomms12964 (cit. on p. 2).

[20] M. H. Devoret and R. J. Schoelkopf: *Superconducting ciruits for quantum information: an outlook.* Science **339**.6124 (2013), pp. 1169–1174. DOI: 10.1126/science.1231930 (cit. on p. 2).

[21] E. Dennis, A. Kitaev, A. Landahl, and J. Preskill: *Topological quantum memory.* Journal of Mathematical Physics **43**.9 (2002), pp. 4452–4505. DOI: 10.1063/1.1499754 (cit. on p. 2).

[22] A. Kitaev: *Fault-tolerant quantum computation by anyons.* Annals of Physics **303**.1 (2003), pp. 2–30. DOI: 10.1016/s0003-4916(02)00018-0 (cit. on p. 2).

[23] A. G. Fowler, M. Mariantoni, J. M. Martinis, and A. N. Cleland: *Surface codes: Towards practical large-scale quantum computation.* Physical Review A **86**.3 (2012). DOI: 10.1103/physreva.86.032324 (cit. on p. 2).

[24] M. D. Reed, L. DiCarlo, S. E. Nigg, L. Sun, L. Frunzio, S. M. Girvin, and R. J. Schoelkopf: *Realization of three-qubit quantum error correction with superconducting circuits*. Nature **482**.7385 (2012), pp. 382–385. DOI: 10.1038/nature10786 (cit. on p. 2).

[25] R. Barends, J. Kelly, A. Megrant, A. Veitia, D. Sank, E. Jeffrey, T. C. White, J. Mutus, A. G. Fowler, B. Campbell, Y. Chen, B. Chiaro, A. Dunsworth, C. Neill, P. O'Malley, P. Roushan, A. Vainsencher, J. Wenner, A. N. Korotkov, A. N. Cleland, and J. M. Martinis: *Superconducting quantum circuits at the surface code threshold for fault tolerance*. Nature **508**.7497 (2014), pp. 500–503. DOI: 10.1038/nature13171 (cit. on p. 2).

[26] J. Kelly, R. Barends, A. G. Fowler, A. Megrant, E. Jeffrey, T. C. White, D. Sank, J. Y. Mutus, B. Campbell, Y. Chen, Z. Chen, B. Chiaro, A. Dunsworth, I.-C. Hoi, C. Neill, P. J. J. O'Malley, C. Quintana, P. Roushan, A. Vainsencher, J. Wenner, A. N. Cleland, and J. M. Martinis: *State preservation by repetitive error detection in a superconducting quantum circuit*. Nature **519**.7541 (2015), pp. 66–69. DOI: 10.1038/nature14270 (cit. on pp. 2, 24).

[27] A. Córcoles, E. Magesan, S. J. Srinivasan, A. W. Cross, M. Steffen, J. M. Gambetta, and J. M. Chow: *Demonstration of a quantum error detection code using a square lattice of four superconducting qubits*. Nature Communications **6**.1 (2015). DOI: 10.1038/ncomms7979 (cit. on p. 2).

[28] S. Bravyi, M. Englbrecht, R. König, and N. Peard: *Correcting coherent errors with surface codes*. npj Quantum Information **4**.1 (2018). DOI: 10.1038/s41534-018-0106-y (cit. on p. 2).

[29] C. L. Degen, F. Reinhard, and P. Cappellaro: *Quantum sensing*. Reviews of Modern Physics **89**.3 (2017). DOI: 10.1103/revmodphys.89.035002 (cit. on pp. 3, 60, 61, 84).

[30] J. Bylander, S. Gustavsson, F. Yan, F. Yoshihara, K. Harrabi, G. Fitch, D. G. Cory, Y. Nakamura, J.-S. Tsai, and W. D. Oliver: *Noise spectroscopy through dynamical decoupling with a superconducting flux qubit*. Nature Physics **7**.7 (2011), pp. 565–570. DOI: 10.1038/nphys1994 (cit. on pp. 3, 61).

[31] S. Schlör, J. Lisenfeld, C. Müller, A. Bilmes, A. Schneider, D. P. Pappas, A. V. Ustinov, and M. Weides: *Correlating Decoherence in Transmon Qubits: Low Frequency Noise by Single Fluctuators*. Physical Review Letters **123**.19 (2019). DOI: 10.1103/physrevlett.123.190502 (cit. on pp. 3, 27, 97).

[32] R. C. Bialczak, R. McDermott, M. Ansmann, M. Hofheinz, N. Katz, E. Lucero, M. Neeley, A. D. O'Connell, H. Wang, A. N. Cleland, and J. M. Martinis: *1/fFlux Noise in Josephson Phase Qubits*. Physical Review Letters **99**.18 (2007). DOI: 10.1103/physrevlett.99.187006 (cit. on p. 3).

[33] F. Yan, J. Bylander, S. Gustavsson, F. Yoshihara, K. Harrabi, D. G. Cory, T. P. Orlando, Y. Nakamura, J.-S. Tsai, and W. D. Oliver: *Spectroscopy of low-frequency noise and its temperature dependence in a superconducting qubit*. Physical Review B **85**.17 (2012). DOI: 10.1103/physrevb.85.174521 (cit. on p. 3).

[34] J. Lisenfeld, G. J. Grabovskij, C. Müller, J. H. Cole, G. Weiss, and A. V. Ustinov: *Observation of directly interacting coherent two-level systems in an amorphous material*. Nature Communications **6**.1 (2015). DOI: 10.1038/ncomms7182 (cit. on pp. 3, 26).

[35] C. Müller, J. Lisenfeld, A. Shnirman, and S. Poletto: *Interacting two-level defects as sources of fluctuating high-frequency noise in superconducting circuits*. Physical Review B **92**.3 (2015), p. 035442. DOI: 10.1103/physrevb.92.035442 (cit. on pp. 3, 26).

[36] J. Lisenfeld, A. Bilmes, S. Matityahu, S. Zanker, M. Marthaler, M. Schechter, G. Schön, A. Shnirman, G. Weiss, and A. V. Ustinov: *Decoherence spectroscopy with individual two-level tunneling defects*. Scientific Reports **6**.1 (2016). DOI: 10.1038/srep23786 (cit. on pp. 3, 26, 61).

[37] J. Lisenfeld, A. Bilmes, A. Megrant, R. Barends, J. Kelly, P. Klimov, G. Weiss, J. M. Martinis, and A. V. Ustinov: *Electric field spectroscopy of material defects in transmon qubits*. npj Quantum Information **5**.1 (2019). DOI: 10.1038/s41534-019-0224-1 (cit. on pp. 3, 120).

[38] Y. Kubo, F. R. Ong, P. Bertet, D. Vion, V. Jacques, D. Zheng, A. Dréau, J.-F. Roch, A. Auffeves, F. Jelezko, J. Wrachtrup, M. F. Barthe, P. Bergonzo, and D. Esteve: *Strong Coupling of a Spin Ensemble to a Superconducting Resonator*. Physical Review Letters **105**.14 (2010). DOI: 10.1103/physrevlett.105.140502 (cit. on p. 3).

[39] S. Probst, H. Rotzinger, S. Wünsch, P. Jung, M. Jerger, M. Siegel, A. V. Ustinov, and P. A. Bushev: *Anisotropic Rare-Earth Spin Ensemble Strongly Coupled to a Superconducting Resonator*. Physical Review Letters **110**.15 (2013). DOI: 10.1103/physrevlett.110.157001 (cit. on p. 3).

[40] S. Probst, N. Kukharchyk, H. Rotzinger, A. Tkalčec, S. Wünsch, A. D. Wieck, M. Siegel, A. V. Ustinov, and P. A. Bushev: *Hybrid quantum circuit with implanted erbium ions*. Applied Physics Letters **105**.16 (2014), p. 162404. DOI: 10.1063/1.4898696 (cit. on p. 3).

[41] S. Probst, A. Tkalčec, H. Rotzinger, D. Rieger, J.-M. L. Floch, M. Goryachev, M. E. Tobar, A. V. Ustinov, and P. A. Bushev: *Three-dimensional cavity quantum electrodynamics with a rare-earth spin ensemble*. Physical Review B **90**.10 (2014). DOI: 10.1103/physrevb.90.100404 (cit. on p. 3).

[42] C. O'Brien, N. Lauk, S. Blum, G. Morigi, and M. Fleischhauer: *Interfacing Superconducting Qubits and Telecom Photons via a Rare-Earth-Doped Crystal.* Physical Review Letters **113**.6 (2014). DOI: 10.1103/physrevlett.113.063603 (cit. on p. 3).

[43] M. Poot and H. S. van der Zant: *Mechanical systems in the quantum regime.* Physics Reports **511**.5 (2012), pp. 273–335. DOI: 10.1016/j.physrep.2011.12.004 (cit. on p. 3).

[44] M. Aspelmeyer, T. J. Kippenberg, and F. Marquardt: *Cavity optomechanics.* Reviews of Modern Physics **86**.4 (2014), pp. 1391–1452. DOI: 10.1103/revmodphys.86.1391 (cit. on p. 3).

[45] R. W. Andrews, R. W. Peterson, T. P. Purdy, K. Cicak, R. W. Simmonds, C. A. Regal, and K. W. Lehnert: *Bidirectional and efficient conversion between microwave and optical light.* Nature Physics **10**.4 (2014), pp. 321–326. DOI: 10.1038/nphys2911 (cit. on p. 3).

[46] R. Hisatomi, A. Osada, Y. Tabuchi, T. Ishikawa, A. Noguchi, R. Yamazaki, K. Usami, and Y. Nakamura: *Bidirectional conversion between microwave and light via ferromagnetic magnons.* Physical Review B **93**.17 (2016). DOI: 10.1103/physrevb.93.174427 (cit. on pp. 3, 102).

[47] C. Kurtsiefer, P. Zarda, M. Halder, H. Weinfurter, P. M. Gorman, P. R. Tapster, and J. G. Rarity: *A step towards global key distribution.* Nature **419**.6906 (2002), pp. 450–450. DOI: 10.1038/419450a (cit. on p. 4).

[48] D. Stucki, N. Gisin, O. Guinnard, G. Ribordy, and H. Zbinden: *Quantum key distribution over 67 km with a plug&play system.* New Journal of Physics **4** (2002), pp. 41–41. DOI: 10.1088/1367-2630/4/1/341 (cit. on p. 4).

[49] G. Kurizki, P. Bertet, Y. Kubo, K. Mølmer, D. Petrosyan, P. Rabl, and J. Schmiedmayer: *Quantum technologies with hybrid systems.* Proceedings of the National Academy of Sciences **112**.13 (2015), pp. 3866–3873. DOI: 10.1073/pnas.1419326112 (cit. on p. 4).

[50] A. V. Chumak, V. I. Vasyuchka, A. A. Serga, and B. Hillebrands: *Magnon spintronics.* Nature Physics **11**.6 (2015), pp. 453–461. DOI: 10.1038/nphys3347 (cit. on p. 4).

[51] B. Lenk, H. Ulrichs, F. Garbs, and M. Münzenberg: *The building blocks of magnonics.* Physics Reports **507**.4-5 (2011), pp. 107–136. DOI: 10.1016/j.physrep.2011.06.003 (cit. on p. 4).

[52] A. V. Chumak, A. A. Serga, and B. Hillebrands: *Magnon transistor for all-magnon data processing.* Nature Communications **5**.1 (2014). DOI: 10.1038/ncomms5700 (cit. on pp. 4, 101).

[53] T. Schneider, A. A. Serga, B. Leven, B. Hillebrands, R. L. Stamps, and M. P. Kostylev: *Realization of spin-wave logic gates*. Applied Physics Letters **92**.2 (2008), p. 022505. DOI: 10.1063/1.2834714 (cit. on p. 4).

[54] S. Klingler, P. Pirro, T. Brächer, B. Leven, B. Hillebrands, and A. V. Chumak: *Design of a spin-wave majority gate employing mode selection*. Applied Physics Letters **105**.15 (2014), p. 152410. DOI: 10.1063/1.4898042 (cit. on p. 4).

[55] V. Cherepanov, I. Kolokolov, and V. L'vov: *The saga of YIG: Spectra, thermodynamics, interaction and relaxation of magnons in a complex magnet*. Physics Reports **229**.3 (1993), pp. 81–144. DOI: 10.1016/0370-1573(93)90107-o (cit. on p. 4).

[56] A. Khitun: *Multi-frequency magnonic logic circuits for parallel data processing*. Journal of Applied Physics **111**.5 (2012), p. 054307. DOI: 10.1063/1.3689011 (cit. on p. 4).

[57] B. A. Kalinikos, N. G. Kovshikov, and A. N. Slavin: *Experimental observation of magnetostatic wave envelope solitons in yttrium iron garnet films*. Physical Review B **42**.13 (1990), pp. 8658–8660. DOI: 10.1103/physrevb.42.8658 (cit. on p. 4).

[58] S. O. Demokritov, V. E. Demidov, O. Dzyapko, G. A. Melkov, A. A. Serga, B. Hillebrands, and A. N. Slavin: *Bose–Einstein condensation of quasi-equilibrium magnons at room temperature under pumping*. Nature **443**.7110 (2006), pp. 430–433. DOI: 10.1038/nature05117 (cit. on p. 4).

[59] A. A. Serga, V. S. Tiberkevich, C. W. Sandweg, V. I. Vasyuchka, D. A. Bozhko, A. V. Chumak, T. Neumann, B. Obry, G. A. Melkov, A. N. Slavin, and B. Hillebrands: *Bose–Einstein condensation in an ultra-hot gas of pumped magnons*. Nature Communications **5**.1 (2014). DOI: 10.1038/ncomms4452 (cit. on p. 4).

[60] F. Bloch: *Zur Theorie des Ferromagnetismus*. Zeitschrift für Physik **61**.3-4 (1930), pp. 206–219. DOI: 10.1007/bf01339661 (cit. on p. 4).

[61] D. D. Stancil and A. Prabhakar: *Spin waves: theory and applications*. New York: Springer, 2009 (cit. on pp. 4, 102, 104).

[62] I. Boventer, M. Pfirrmann, J. Krause, Y. Schön, M. Kläui, and M. Weides: *Complex temperature dependence of coupling and dissipation of cavity magnon polaritons from millikelvin to room temperature*. Physical Review B **97**.18 (2018). DOI: 10.1103/physrevb.97.184420 (cit. on pp. 5, 106, 110, 121).

[63] A. F. van Loo, R. G. E. Morris, and A. D. Karenowska: *Time-Resolved Measurements of Surface Spin-Wave Pulses at Millikelvin Temperatures*. Physical Review Applied **10**.4 (2018). DOI: 10.1103/physrevapplied.10.044070 (cit. on p. 5).

[64] M. Pfirrmann, I. Boventer, A. Schneider, T. Wolz, M. Kläui, A. V. Ustinov, and M. Weides: *Magnons at low excitations: Observation of incoherent coupling to a bath of two-level systems*. Physical Review Research **1**.3 (2019), p. 032023. DOI: 10.1103/PhysRevResearch.1.032023 (cit. on pp. 5, 38, 84, 106, 121).

[65] S. Kosen, A. F. van Loo, D. A. Bozhko, L. Mihalceanu, and A. D. Karenowska: *Microwave magnon damping in YIG films at millikelvin temperatures*. APL Materials **7**.10 (2019), p. 101120. DOI: 10.1063/1.5115266 (cit. on pp. 5, 107).

[66] Y. Tabuchi, S. Ishino, A. Noguchi, T. Ishikawa, R. Yamazaki, K. Usami, and Y. Nakamura: *Coherent coupling between a ferromagnetic magnon and a superconducting qubit*. Science **349**.6246 (2015), pp. 405–408. DOI: 10.1126/science.aaa3693 (cit. on pp. 5, 84, 121).

[67] Y. Tabuchi, S. Ishino, A. Noguchi, T. Ishikawa, R. Yamazaki, K. Usami, and Y. Nakamura: *Quantum magnonics: The magnon meets the superconducting qubit*. Comptes Rendus Physique **17**.7 (2016), pp. 729–739. DOI: 10.1016/j.crhy.2016.07.009 (cit. on pp. 5, 106).

[68] D. Lachance-Quirion, Y. Tabuchi, S. Ishino, A. Noguchi, T. Ishikawa, R. Yamazaki, and Y. Nakamura: *Resolving quanta of collective spin excitations in a millimeter-sized ferromagnet*. Science Advances **3**.7 (2017), e1603150. DOI: 10.1126/sciadv.1603150 (cit. on pp. 5, 106).

[69] D. Lachance-Quirion, S. P. Wolski, Y. Tabuchi, S. Kono, K. Usami, and Y. Nakamura: *Entanglement-based single-shot detection of a single magnon with a superconducting qubit*. arXiv (2019) (cit. on pp. 5, 106, 121).

[70] A. Schneider, T. Wolz, M. Pfirrmann, M. Spiecker, H. Rotzinger, A. V. Ustinov, and M. Weides: *Transmon qubit in a magnetic field: Evolution of coherence and transition frequency*. Physical Review Research **1**.2 (2019). DOI: 10.1103/physrevresearch.1.023003 (cit. on pp. 5, 6, 83, 85, 86, 88, 90, 91, 93, 96–98, 120).

[71] A. Schneider, J. Braumüller, L. Guo, P. Stehle, H. Rotzinger, M. Marthaler, A. V. Ustinov, and M. Weides: *Local sensing with the multilevel ac Stark effect*. Physical Review A **97**.6 (2018). DOI: 10.1103/PhysRevA.97.062334 (cit. on pp. 6, 56, 59, 63, 69, 70, 72, 75, 76, 78, 80, 83, 119).

[72] M. Kristen, A. Schneider, A. Stehli, T. Wolz, S. Danilin, H. S. Ku, J. Long, X. Wu, R. Lake, D. P. Pappas, A. V. Ustinov, and M. Weides: *Amplitude and frequency sensing of microwave fields with a superconducting transmon qudit*. npj Quantum Information **6**.1 (2020). DOI: 10.1038/s41534-020-00287-w (cit. on pp. 6, 79, 80, 82, 119).

[73] J. J. Sakurai: *Modern Quantum Mechanics, Revised Edition*. Ed. by S. F. Tuan. Addison-Wesley Publishing Company, 1994 (cit. on p. 11).

[74] J. R. Friedman, V. Patel, W. Chen, S. K. Tolpygo, and J. E. Lukens: *Quantum superposition of distinct macroscopic states*. Nature **406**.6791 (2000), pp. 43–46. DOI: 10.1038/35017505 (cit. on p. 11).

[75] E. Jaynes and F. Cummings: *Comparison of quantum and semiclassical radiation theories with application to the beam maser*. Proceedings of the IEEE **51**.1 (1963), pp. 89–109. DOI: 10.1109/proc.1963.1664 (cit. on p. 11).

[76] A. Blais, R.-S. Huang, A. Wallraff, S. M. Girvin, and R. J. Schoelkopf: *Cavity quantum electrodynamics for superconducting electrical circuits: An architecture for quantum computation*. Physical Review A **69**.6 (2004), p. 062320. DOI: 10.1103/physreva.69.062320 (cit. on pp. 11, 62, 63).

[77] A. Schneider: *Quantum state tomography and benchmarking on superconducting qubits*. MA thesis. Karlsruhe Institute of Technology, 2015 (cit. on pp. 11, 25, 48, 50, 53).

[78] F. Jelezko, T. Gaebel, I. Popa, A. Gruber, and J. Wrachtrup: *Observation of Coherent Oscillations in a Single Electron Spin*. Physical Review Letters **92**.7 (2004). DOI: 10.1103/physrevlett.92.076401 (cit. on p. 12).

[79] E. Farhi and S. Gutmann: *Analog analogue of a digital quantum computation*. Physical Review A **57**.4 (1998), pp. 2403–2406. DOI: 10.1103/physreva.57.2403 (cit. on p. 12).

[80] C. Godfrin, A. Ferhat, R. Ballou, S. Klyatskaya, M. Ruben, W. Wernsdorfer, and F. Balestro: *Operating Quantum States in Single Magnetic Molecules: Implementation of Grover's Quantum Algorithm*. Physical Review Letters **119**.18 (2017). DOI: 10.1103/physrevlett.119.187702 (cit. on p. 12).

[81] H. Kammerlingh-Onnes: *The resistance of pure mercury at helium temperatures*. Commun. Phys. Lab. Univ. Leiden **12** (1911), p. 120 (cit. on p. 13).

[82] W. Meissner and R. Ochsenfeld: *Ein neuer Effekt bei Eintritt der Supraleitfähigkeit*. Naturwissenschaften **21**.44 (1933), pp. 787–788. DOI: 10.1007/BF01504252 (cit. on p. 13).

[83] J. Bardeen, L. N. Cooper, and J. R. Schrieffer: *Theory of Superconductivity*. Physical Review **108**.5 (1957). DOI: 10.1103/PhysRev.108.1175 (cit. on p. 13).

[84] W. Pauli: *Über den Zusammenhang des Abschlusses der Elektronengruppen im Atom mit der Komplexstruktur der Spektren*. Zeitschrift für Physik **31**.1 (1925), pp. 765–783. DOI: 10.1007/bf02980631 (cit. on p. 14).

[85] V. L. Ginzburg and L. D. Landau: *To the Theory of Superconductivity*. Zh. Eksp. Teor. Fiz **20**.1064 (1950) (cit. on pp. 16, 17).

[86] M. Tinkham: *Introduction to Superconductivity, 2nd edition*. Dover, 2004 (cit. on pp. 16, 17, 27).

[87] W. Buckel and R. Kleiner: *Supraleitung*. Wiley-VCH, 2012. DOI: 10.1002/9783527668670 (cit. on pp. 16, 17).

[88] V. V. Schmidt: *The Physics of Superconductors*. Ed. by P. Müller and A. Ustinov. Springer, 1997 (cit. on pp. 16, 17).

[89] A. M. Toxen: *Critical Fields of Thin Superconducting Films. I. Thickness Effects*. Physical Review **127**.2 (1962), pp. 382–386. DOI: 10.1103/physrev.127.382 (cit. on p. 17).

[90] B. L. Brandt, R. D. Parks, and R. D. Chaudhari: *Intermediate state of thin superconductors*. Journal of Low Temperature Physics **4**.1 (1971), pp. 41–63. DOI: 10.1007/bf00628436 (cit. on p. 18).

[91] R. Meservey and B. B. Schwartz: *Superconductivity: In Two Volumes: Volume 2 (Part 2)*. Ed. by R. D. Parks. CRC Press, 1969 (cit. on p. 18).

[92] B. D. Josephson: *Possible new effects in superconductive tunnelling*. Physics Letters **1**.7 (1962), pp. 251–253. DOI: 10.1016/0031-9163(62)91369-0 (cit. on p. 18).

[93] A. Barone and G. Paterno, eds.: *Physics and applications of the Josephson effect*. New York: Wiley, 1982. URL: http://onlinelibrary.wiley.com/book/10.1002/352760278X (cit. on pp. 20, 27).

[94] M. H. Devoret: *Quantum fluctuations. Les Houches, Session LXIII*. Ed. by E. G. S. Reynaud and J. Zinn-Justin. Elsevier Science B. V., 1997. Chap. Quantum fluctuations in electrical circuits, pp. 351–386. URL: http://qulab.eng.yale.edu/documents/reprints/Houches_fluctuations.pdf (cit. on p. 21).

[95] B. Yurke and J. S. Denker: *Quantum network theory*. Physical Review A **29**.3 (1984), pp. 1419–1437. DOI: 10.1103/physreva.29.1419 (cit. on p. 21).

[96] L. S. Bishop: *Circuit Quantum Electrodynamics*. PhD thesis. Yale University, 2010. URL: http://arxiv.org/abs/1007.3520 (cit. on p. 22).

[97] J. Braumüller, J. Cramer, S. Schlör, H. Rotzinger, L. Radtke, A. Lukashenko, P. Yang, S. T. Skacel, S. Probst, M. Marthaler, L. Guo, A. V. Ustinov, and M. Weides: *Multiphoton dressing of an anharmonic superconducting many-level quantum circuit*. Physical Review B **91**.5 (2015), p. 054523. DOI: 10.1103/physrevb.91.054523 (cit. on pp. 22, 62).

[98] M. Sandberg, M. R. Vissers, T. A. Ohki, J. Gao, J. Aumentado, M. Weides, and D. P. Pappas: *Radiation-suppressed superconducting quantum bit in a planar geometry*. Applied Physics Letters **102**.7 (2013), p. 072601. DOI: 10.1063/1.4792698 (cit. on pp. 24, 25).

[99] J. Braumüller, M. Sandberg, M. R. Vissers, A. Schneider, S. Schlör, L. Grünhaupt, H. Rotzinger, M. Marthaler, A. Lukashenko, A. Dieter, A. V. Ustinov, M. Weides, and D. P. Pappas: *Concentric transmon qubit featuring fast tunability and an anisotropic magnetic dipole moment*. Applied Physics Letters **108**.3 (2016), p. 032601. DOI: 10.1063/1.4940230 (cit. on pp. 24, 25, 56).

[100] J. Rahamim, T. Behrle, M. J. Peterer, A. Patterson, P. A. Spring, T. Tsunoda, R. Manenti, G. Tancredi, and P. J. Leek: *Double-sided coaxial circuit QED with out-of-plane wiring*. Applied Physics Letters **110**.22 (2017), p. 222602. DOI: 10.1063/1.4984299 (cit. on p. 24).

[101] R. Barends, J. Kelly, A. Megrant, D. Sank, E. Jeffrey, Y. Chen, Y. Yin, B. Chiaro, J. Mutus, C. Neill, P. O'Malley, P. Roushan, J. Wenner, T. White, A. N. Cleland, and J. M. Martinis: *Coherent Josephson Qubit Suitable for Scalable Quantum Integrated Circuits*. Physical Review Letters **111**.8 (2013), p. 080502. DOI: 10.1103/physrevlett.111.080502 (cit. on p. 24).

[102] J. C. Braumüller: *Quantum simulation experiments with superconducting circuits*. PhD thesis. Karlsruhe Institute of Technology, 2018. DOI: 10.5445/KSP/1000081315 (cit. on pp. 25, 26, 111).

[103] H. Paik, D. I. Schuster, L. S. Bishop, G. Kirchmair, G. Catelani, A. P. Sears, B. R. Johnson, M. J. Reagor, L. Frunzio, L. I. Glazman, S. M. Girvin, M. H. Devoret, and R. J. Schoelkopf: *Observation of High Coherence in Josephson Junction Qubits Measured in a Three-Dimensional Circuit QED Architecture*. Physical Review Letters **107**.24 (2011). DOI: 10.1103/PhysRevLett.107.240501 (cit. on pp. 26, 27).

[104] M. Sandberg, M. R. Vissers, J. S. Kline, M. Weides, J. Gao, D. S. Wisbey, and D. P. Pappas: *Etch induced microwave losses in titanium nitride superconducting resonators*. Applied Physics Letters **100**.26 (2012), p. 262605. DOI: 10.1063/1.4729623 (cit. on p. 26).

[105] J. M. Martinis, K. B. Cooper, R. McDermott, M. Steffen, M. Ansmann, K. D. Osborn, K. Cicak, S. Oh, D. P. Pappas, R. W. Simmonds, and C. Y. Clare: *Decoherence in Josephson Qubits from Dielectric Loss*. Physical Review Letters **95**.21 (2005), p. 210503. DOI: 10.1103/physrevlett.95.210503 (cit. on p. 26).

[106] G. J. Grabovskij, T. Peichl, J. Lisenfeld, G. Weiss, and A. V. Ustinov: *Strain Tuning of Individual Atomic Tunneling Systems Detected by a Superconducting Qubit*. Science **338**.6104 (2012), pp. 232–234. DOI: 10.1126/science.1226487 (cit. on p. 26).

[107] H. Paik and K. D. Osborn: *Reducing quantum-regime dielectric loss of silicon nitride for superconducting quantum circuits*. Applied Physics Letters **96**.7 (2010), p. 072505. DOI: 10.1063/1.3309703 (cit. on p. 26).

[108] G. Catelani, J. Koch, L. Frunzio, R. J. Schoelkopf, M. H. Devoret, and L. I. Glazman: *Quasiparticle Relaxation of Superconducting Qubits in the Presence of Flux*. Physical Review Letters **106**.7 (2011). DOI: 10.1103/physrevlett.106.077002 (cit. on p. 27).

[109] J. M. Martinis, M. Ansmann, and J. Aumentado: *Energy Decay in Superconducting Josephson-Junction Qubits from Nonequilibrium Quasiparticle Excitations*. Physical Review Letters **103**.9 (2009). DOI: 10.1103/physrevlett.103.097002 (cit. on p. 27).

[110] D. Ristè, C. C. Bultink, M. J. Tiggelman, R. N. Schouten, K. W. Lehnert, and L. DiCarlo: *Millisecond charge-parity fluctuations and induced decoherence in a superconducting transmon qubit*. Nature Communications **4** (2013), p. 1913. DOI: 10.1038/ncomms2936 (cit. on p. 27).

[111] A. Ferguson, N. Court, F. Hudson, and R. Clark: *Thin-film aluminium for superconducting qubits. 2006 International Conference on Nanoscience and Nanotechnology*. IEEE, 2006. DOI: 10.1109/iconn.2006.340682 (cit. on p. 27).

[112] L. Grünhaupt, N. Maleeva, S. T. Skacel, M. Calvo, F. Levy-Bertrand, A. V. Ustinov, H. Rotzinger, A. Monfardini, G. Catelani, and I. M. Pop: *Loss Mechanisms and Quasiparticle Dynamics in Superconducting Microwave Resonators Made of Thin-Film Granular Aluminum*. Physical Review Letters **121**.11 (2018). DOI: 10.1103/physrevlett.121.117001 (cit. on p. 28).

[113] C. Wang, Y. Y. Gao, I. M. Pop, U. Vool, C. Axline, T. Brecht, R. W. Heeres, L. Frunzio, M. H. Devoret, G. Catelani, L. I. Glazman, and R. J. Schoelkopf: *Measurement and control of quasiparticle dynamics in a superconducting qubit*. Nature Communications **5**.1 (2014). DOI: 10.1038/ncomms6836 (cit. on pp. 28, 84, 95).

[114] U. Vool, I. Pop, K. Sliwa, B. Abdo, C. Wang, T. Brecht, Y. Gao, S. Shankar, M. Hatridge, G. Catelani, M. Mirrahimi, L. Frunzio, R. Schoelkopf, L. Glazman, and M. Devoret: *Non-Poissonian Quantum Jumps of a Fluxonium Qubit due to Quasiparticle Excitations*. Physical Review Letters **113**.24 (2014). DOI: 10.1103/physrevlett.113.247001 (cit. on p. 28).

[115] S. Kwon, A. F. Roudsari, O. W. B. Benningshof, Y.-C. Tang, H. R. Mohebbi, I. A. J. Taminiau, D. Langenberg, S. Lee, G. Nichols, D. G. Cory, and G.-X. Miao: *Magnetic field dependent microwave losses in superconducting niobium microstrip resonators*. Journal of Applied Physics **124**.3 (2018), p. 033903. DOI: 10.1063/1.5027003 (cit. on pp. 28, 90, 95).

[116] X. Xi, J. Hwang, C. Martin, D. H. Reitze, C. J. Stanton, D. B. Tanner, and G. L. Carr: *Effect of a magnetic field on the quasiparticle recombination in superconductors*. Physical Review B **87**.14 (2013). DOI: 10.1103/physrevb.87.140502 (cit. on pp. 28, 90, 95).

[117] J. Aumentado, M. W. Keller, J. M. Martinis, and M. H. Devoret: *Nonequilibrium Quasiparticles and 2e Periodicity in Single-Cooper-Pair Transistors*. Physical Review Letters **92**.6 (2004). DOI: 10.1103/physrevlett.92.066802 (cit. on p. 28).

[118] L. Sun, L. DiCarlo, M.D. Reed, G. Catelani, L.S. Bishop, D.I. Schuster, B.R. Johnson, G.A. Yang, L. Frunzio, L. Glazman, M.H. Devoret, and R.J. Schoelkopf: *Measurements of Quasiparticle Tunneling Dynamics in a Band-Gap-Engineered Transmon Qubit*. Physical Review Letters **108**.23 (2012). DOI: 10.1103/physrevlett.108.230509 (cit. on p. 28).

[119] S. Rajauria, L.M.A. Pascal, P. Gandit, F.W.J. Hekking, B. Pannetier, and H. Courtois: *Efficiency of quasiparticle evacuation in superconducting devices*. Physical Review B **85**.2 (2012). DOI: 10.1103/physrevb.85.020505 (cit. on p. 28).

[120] R.-P. Riwar, A. Hosseinkhani, L.D. Burkhart, Y.Y. Gao, R.J. Schoelkopf, L.I. Glazman, and G. Catelani: *Normal-metal quasiparticle traps for superconducting qubits*. Physical Review B **94**.10 (2016). DOI: 10.1103/physrevb.94.104516 (cit. on p. 28).

[121] A. Hosseinkhani, R.-P. Riwar, R.J. Schoelkopf, L.I. Glazman, and G. Catelani: *Optimal Configurations for Normal-Metal Traps in Transmon Qubits*. Physical Review Applied **8**.6 (2017). DOI: 10.1103/physrevapplied.8.064028 (cit. on p. 28).

[122] I. Nsanzineza and B. Plourde: *Trapping a Single Vortex and Reducing Quasiparticles in a Superconducting Resonator*. Physical Review Letters **113**.11 (2014). DOI: 10.1103/physrevlett.113.117002 (cit. on p. 28).

[123] M. Taupin, I.M. Khaymovich, M. Meschke, A.S. Mel'nikov, and J.P. Pekola: *Tunable quasiparticle trapping in Meissner and vortex states of mesoscopic superconductors*. Nature Communications **7**.1 (2016). DOI: 10.1038/ncomms10977 (cit. on p. 28).

[124] R. Benaroya and K.W. Shepard: *Superconducting rf cavity in a dc magnetic field*. Review of Scientific Instruments **59**.9 (1988), pp. 2100–2101. DOI: 10.1063/1.1140038 (cit. on p. 28).

[125] E. Silva and N. Pompeo: *A short note on flux lines and flux motion*. Lecture Note. Università Roma Tre, 2012. URL: http://www.sea.uniroma3.it/eldem/disp/FluxMatter.pdf (cit. on p. 29).

[126] J. Bardeen and M.J. Stephen: *Theory of the Motion of Vortices in Superconductors*. Physical Review **140**.4A (1965), A1197–A1207. DOI: 10.1103/physrev.140.a1197 (cit. on p. 29).

[127] C.F. Hempstead and Y.B. Kim: *Resistive Transitions and Surface effects in Type-II Superconductors*. Physical Review Letters **12**.6 (1964), pp. 145–148. DOI: 10.1103/physrevlett.12.145 (cit. on p. 29).

[128] A.R. Strnad, C.F. Hempstead, and Y.B. Kim: *Dissipative Mechanism in Type-II Superconductors*. Physical Review Letters **13**.26 (1964), pp. 794–797. DOI: 10.1103/physrevlett.13.794 (cit. on p. 29).

[129] C. Song, T. W. Heitmann, M. P. DeFeo, K. Yu, R. McDermott, M. Neeley, J. M. Martinis, and B. L. T. Plourde: *Microwave response of vortices in superconducting thin films of Re and Al*. Physical Review B **79**.17 (2009). DOI: 10.1103/physrevb.79.174512 (cit. on p. 29).

[130] D. Bothner, T. Gaber, M. Kemmler, D. Koelle, and R. Kleiner: *Improving the performance of superconducting microwave resonators in magnetic fields*. Applied Physics Letters **98**.10 (2011), p. 102504. DOI: 10.1063/1.3560480 (cit. on pp. 29, 90).

[131] D. Bothner, T. Gaber, M. Kemmler, D. Koelle, R. Kleiner, S. Wünsch, and M. Siegel: *Magnetic hysteresis effects in superconducting coplanar microwave resonators*. Physical Review B **86**.1 (2012). DOI: 10.1103/physrevb.86.014517 (cit. on pp. 29, 30, 90, 91).

[132] C. P. Bean: *Magnetization of Hard Superconductors*. Physical Review Letters **8**.6 (1962), pp. 250–253. DOI: 10.1103/physrevlett.8.250 (cit. on pp. 29, 90, 94).

[133] C. P. Bean: *Magnetization of High-Field Superconductors*. Reviews of Modern Physics **36**.1 (1964), pp. 31–39. DOI: 10.1103/revmodphys.36.31 (cit. on pp. 29, 90, 94).

[134] C. Jooss, J. Albrecht, H. Kuhn, S. Leonhardt, and H. Kronmüller: *Magneto-optical studies of current distributions in high-Tcsuperconductors*. Reports on Progress in Physics **65**.5 (2002), pp. 651–788. DOI: 10.1088/0034-4885/65/5/202 (cit. on p. 30).

[135] D. M. Pozar: *Microwave Engineering*. 4th ed. Wiley, 2012. URL: http://www.worldcat.org/isbn/0470631554 (cit. on pp. 32–34).

[136] S. Probst, F. B. Song, P. A. Bushev, A. V. Ustinov, and M. Weides: *Efficient and robust analysis of complex scattering data under noise in microwave resonators*. Review of Scientific Instruments **86**.2 (2015), p. 024706. DOI: 10.1063/1.4907935 (cit. on pp. 35, 90).

[137] D. F. Walls and G. J. Milburn: *Quantum optics*. 2. ed. Berlin: Springer, 2008 (cit. on p. 38).

[138] T. Wolz, A. Stehli, A. Schneider, I. Boventer, R. Macêdo, A. V. Ustinov, M. Kläui, and M. Weides: *Introducing coherent time control to cavity magnon-polariton modes*. Communications Physics **3**.1 (2020). DOI: 10.1038/s42005-019-0266-x (cit. on p. 40).

[139] D. D. Osheroff, R. C. Richardson, and D. M. Lee: *Evidence for a New Phase of Solid He3*. Physical Review Letters **28**.14 (1972), pp. 885–888. DOI: 10.1103/physrevlett.28.885 (cit. on p. 41).

[140] M. Pfirrmann: *Adding nonlinearity to an electromagnetic-magnonic quantum hybrid device*. PhD thesis. Karlsruhe Institute of Technology, 2019. 120 pp. DOI: 10.5445/IR/1000100336 (cit. on pp. 46, 106).

[141] *Qkit – a quantum measurement and evaluation suite in python*. Karlsruhe Institute of Technology. URL: https://github.com/qkitgroup/qkit (cit. on pp. 53, 148).

[142] M. R. Vissers, J. Gao, J. S. Kline, M. Sandberg, M. P. Weides, D. S. Wisbey, and D. P. Pappas: *Characterization and in-situ monitoring of sub-stoichiometric adjustable superconducting critical temperature titanium nitride growth*. Thin Solid Films **548** (2013), pp. 485–488. DOI: 10.1016/j.tsf.2013.07.046 (cit. on p. 55).

[143] J. B. Chang, M. R. Vissers, A. D. Córcoles, M. Sandberg, J. Gao, D. W. Abraham, J. M. Chow, J. M. Gambetta, M. Beth Rothwell, G. A. Keefe, M. Steffen, and D. P. Pappas: *Improved superconducting qubit coherence using titanium nitride*. Applied Physics Letters **103**.1 (2013), p. 012602. DOI: 10.1063/1.4813269 (cit. on p. 55).

[144] R. C. Jaklevic, J. Lambe, A. H. Silver, and J. E. Mercereau: *Quantum Interference Effects in Josephson Tunneling*. Physical Review Letters **12**.7 (1964), pp. 159–160. DOI: 10.1103/physrevlett.12.159 (cit. on p. 59).

[145] O. Dössel: *Bildgebende Verfahren in der Medizin : Von der Technik zur medizinischen Anwendung*. 2. Aufl. 2016. SpringerLink. Berlin, Heidelberg: Springer Vieweg, 2016 (cit. on p. 60).

[146] D. Vasyukov, Y. Anahory, L. Embon, D. Halbertal, J. Cuppens, L. Neeman, A. Finkler, Y. Segev, Y. Myasoedov, M. L. Rappaport, M. E. Huber, and E. Zeldov: *A scanning superconducting quantum interference device with single electron spin sensitivity*. Nature Nanotechnology **8**.9 (2013), pp. 639–644. DOI: 10.1038/nnano.2013.169 (cit. on p. 61).

[147] D. Halbertal, J. Cuppens, M. B. Shalom, L. Embon, N. Shadmi, Y. Anahory, H. R. Naren, J. Sarkar, A. Uri, Y. Ronen, Y. Myasoedov, L. S. Levitov, E. Joselevich, A. K. Geim, and E. Zeldov: *Nanoscale thermal imaging of dissipation in quantum systems*. Nature **539**.7629 (2016), pp. 407–410. DOI: 10.1038/nature19843 (cit. on p. 61).

[148] O. Astafiev, Y. Pashkin, Y. Nakamura, T. Yamamoto, and J. Tsai: *Quantum Noise in the Josephson Charge Qubit*. Physical Review Letters **93**.26 (2004). DOI: 10.1103/physrevlett.93.267007 (cit. on p. 61).

[149] J. Stark: *Beobachtungen über den Effekt des elektrischen Feldes auf Spektrallinien. V. Feinzerlegung der Wasserstoffserie*. Annalen der Physik **353**.18 (1915), pp. 193–209. DOI: 10.1002/andp.19153531804 (cit. on p. 62).

[150] E. Schrödinger: *Quantisierung als Eigenwertproblem*. Annalen der Physik **385**.13 (1926), pp. 437–490. DOI: 10.1002/andp.19263851302 (cit. on p. 62).

[151] A. M. Bonch-Bruevich, N. N. Kostin, V. A. Khodovoĭ, and V. V. Khromov: *Changes in the Atomic Absorption Spectrum in the Field of a Light Wave. I.* Soviet Journal of Experimental and Theoretical Physics **29** (1969), p. 82. URL: http://www.jetp.ac.ru/cgi-bin/dn/e_029_01_0082.pdf (cit. on p. 62).

[152] D. I. Schuster, A. Wallraff, A. Blais, L. Frunzio, R.-S. Huang, J. Majer, S. M. Girvin, and R. J. Schoelkopf: *ac Stark Shift and Dephasing of a Superconducting Qubit Strongly Coupled to a Cavity Field.* Physical Review Letters **94**.12 (2005). DOI: 10.1103/physrevlett.94.123602 (cit. on p. 62).

[153] N. B. Delone and V. P. Krainov: *AC Stark shift of atomic energy levels.* Physics-Uspekhi **42**.7 (1999), pp. 669–687. DOI: 10.1070/pu1999v042n07abeh000557 (cit. on p. 62).

[154] H. Wang, M. Hofheinz, J. Wenner, M. Ansmann, R. C. Bialczak, M. Lenander, E. Lucero, M. Neeley, A. D. O'Connell, D. Sank, M. Weides, A. N. Cleland, and J. M. Martinis: *Improving the coherence time of superconducting coplanar resonators.* Applied Physics Letters **95**.23 (2009), p. 233508. DOI: 10.1063/1.3273372 (cit. on p. 64).

[155] G. Lindblad: *On the generators of quantum dynamical semigroups.* Communications in Mathematical PHysics **48**.2 (1976), pp. 119–130. DOI: 10.1007/BF01608499 (cit. on p. 70).

[156] S. Haroche and J.-M. Raimond: *Exploring the Quantum - Atoms, Cavities, and Photons.* Oxford University Press, 2006 (cit. on p. 70).

[157] X. Jin, A. Kamal, A. Sears, T. Gudmundsen, D. Hover, J. Miloshi, R. Slattery, F. Yan, J. Yoder, T. Orlando, S. Gustavsson, and W. Oliver: *Thermal and Residual Excited-State Population in a 3D Transmon Qubit.* Physical Review Letters **114**.24 (2015). DOI: 10.1103/physrevlett.114.240501 (cit. on p. 70).

[158] J. Johansson, P. Nation, and F. Nori: *QuTiP: An open-source Python framework for the dynamics of open quantum systems.* Comp. Phys. Comm. **183**.8 (2012), pp. 1760–1772. DOI: 10.1016/j.cpc.2012.02.021 (cit. on p. 71).

[159] J. Johansson, P. Nation, and F. Nori: *QuTiP 2: A Python framework for the dynamics of open quantum systems.* Comp. Phys. Comm. **184**.4 (2013), pp. 1234–1240. DOI: 10.1016/j.cpc.2012.11.019 (cit. on p. 71).

[160] M. Kristen: *Time resolved quantum sensing of microwave fields with a transmon qubit.* MA thesis. Karlsruhe Institue of Technology, 2019 (cit. on pp. 79, 80).

[161] S. Danilin, A. V. Lebedev, A. Vepsäläinen, G. B. Lesovik, G. Blatter, and G. S. Paraoanu: *Quantum-enhanced magnetometry by phase estimation algorithms with a single artificial atom.* npj Quantum Information **4**.1 (2018). DOI: 10.1038/s41534-018-0078-y (cit. on p. 80).

[162] K. Serniak, M. Hays, G. de Lange, S. Diamond, S. Shankar, L. Burkhart, L. Frunzio, M. Houzet, and M. Devoret: *Hot Nonequilibrium Quasiparticles in Transmon Qubits*. Physical Review Letters **121**.15 (2018). DOI: `10.1103/physrevlett.121.157701` (cit. on p. 83).

[163] A. D. Córcoles, J. M. Chow, J. M. Gambetta, C. Rigetti, J. R. Rozen, G. A. Keefe, M. B. Rothwell, M. B. Ketchen, and M. Steffen: *Protecting superconducting qubits from radiation*. Applied Physics Letters **99**.18 (2011), p. 181906. DOI: `10.1063/1.3658630` (cit. on p. 83).

[164] S. Schlör: *Temperature dependence of coherence in transmon qubits*. MA thesis. Karlsruhe Institute of Technology, 2015 (cit. on p. 83).

[165] S. M. Meißner: *Dynamik atomarer Tunnelsysteme in dünnen, ungeordneten AlOx-Schichten - Einzelne Tunnelsysteme in Josephson-Kontakten und Zustandsdichte in Al-AlOx-Al-Kondensatoren*. PhD thesis. Karlsruhe Institute of Technology, 2017. DOI: `10.5445/IR/1000072415` (cit. on p. 84).

[166] R. Barends, J. Wenner, M. Lenander, Y. Chen, R. C. Bialczak, J. Kelly, E. Lucero, P. O'Malley, M. Mariantoni, D. Sank, H. Wang, T. C. White, Y. Yin, J. Zhao, A. N. Cleland, J. M. Martinis, and J. J. A. Baselmans: *Minimizing quasiparticle generation from stray infrared light in superconducting quantum circuits*. Applied Physics Letters **99**.11 (2011), p. 113507. DOI: `10.1063/1.3638063` (cit. on p. 84).

[167] F. Luthi, T. Stavenga, O. W. Enzing, A. Bruno, C. Dickel, N. K. Langford, M. A. Rol, T. S. Jespersen, J. Nygård, P. Krogstrup, and L. DiCarlo: *Evolution of Nanowire Transmon Qubits and Their Coherence in a Magnetic Field*. Physical Review Letters **120**.10 (2018), p. 100502. DOI: `10.1103/physrevlett.120.100502` (cit. on pp. 84, 122).

[168] J. M. Kreikebaum, A. Dove, W. Livingston, E. Kim, and I. Siddiqi: *Optimization of infrared and magnetic shielding of superconducting TiN and Al coplanar microwave resonators*. Superconductor Science and Technology **29**.10 (2016), p. 104002. DOI: `10.1088/0953-2048/29/10/104002` (cit. on p. 84).

[169] D. Flanigan, B. R. Johnson, M. H. Abitbol, S. Bryan, R. Cantor, P. Day, G. Jones, P. Mauskopf, H. McCarrick, A. Miller, and J. Zmuidzinas: *Magnetic field dependence of the internal quality factor and noise performance of lumped-element kinetic inductance detectors*. Applied Physics Letters **109**.14 (2016), p. 143503. DOI: `10.1063/1.4964119` (cit. on p. 84).

[170] V. Mourik, K. Zuo, S. M. Frolov, S. R. Plissard, E. P. A. M. Bakkers, and L. P. Kouwenhoven: *Signatures of Majorana Fermions in Hybrid Superconductor-Semiconductor Nanowire Devices*. Science **336**.6084 (2012), pp. 1003–1007. DOI: `10.1126/science.1222360` (cit. on p. 84).

[171] A. Dunsworth, A. Megrant, C. Quintana, Z. Chen, R. Barends, B. Burkett, B. Foxen, Y. Chen, B. Chiaro, A. Fowler, R. Graff, E. Jeffrey, J. Kelly, E. Lucero, J. Y. Mutus, M. Neeley, C. Neill, P. Roushan, D. Sank, A. Vainsencher, J. Wenner, T. C. White, and J. M. Martinis: *Characterization and reduction of capacitive loss induced by sub-micron Josephson junction fabrication in superconducting qubits.* Applied Physics Letters **111**.2 (2017), p. 022601. DOI: 10.1063/1.4993577 (cit. on p. 89).

[172] X. Wu, J. L. Long, H. S. Ku, R. E. Lake, M. Bal, and D. P. Pappas: *Overlap junctions for high coherence superconducting qubits.* Applied Physics Letters **111**.3 (2017), p. 032602. DOI: 10.1063/1.4993937 (cit. on p. 89).

[173] V. Ambegaokar and A. Baratoff: *Tunneling Between Superconductors.* Physical Review Letters **10**.11 (1963), pp. 486–489. DOI: 10.1103/physrevlett.10.486 (cit. on p. 89).

[174] V. Ambegaokar and A. Baratoff: *Tunneling Between Superconductors (Erratum).* Physical Review Letters **11**.2 (1963), p. 104. DOI: 10.1103/physrevlett.11.104 (cit. on p. 89).

[175] D. Bothner, D. Wiedmaier, B. Ferdinand, R. Kleiner, and D. Koelle: *Improving Superconducting Resonators in Magnetic Fields by Reduced Field Focussing and Engineered Flux Screening.* Physical Review Applied **8**.3 (2017). DOI: 10.1103/physrevapplied.8.034025 (cit. on p. 90).

[176] N. Groll, A. Gurevich, and I. Chiorescu: *Measurement of the nonlinear Meissner effect in superconducting Nb films using a resonant microwave cavity: A probe of unconventional pairing symmetries.* Physical Review B **81**.2 (2010). DOI: 10.1103/physrevb.81.020504 (cit. on p. 94).

[177] G. Ithier, E. Collin, P. Joyez, P. J. Meeson, D. Vion, D. Esteve, F. Chiarello, A. Shnirman, Y. Makhlin, J. Schriefl, and G. Schön: *Decoherence in a superconducting quantum bit circuit.* Physical Review B **72**.13 (2005), p. 134519. DOI: 10.1103/physrevb.72.134519 (cit. on p. 98).

[178] Q. Wang, M. Kewenig, M. Schneider, R. Verba, B. Heinz, M. Geilen, M. Mohseni, B. Lägel, F. Ciubotaru, C. Adelmann, C. Dubs, S. D. Cotofana, T. Brächer, P. Pirro, and A. V. Chumak: *Realization of a nanoscale magnonic directional coupler for all-magnon circuits.* arXiv (2019). URL: http://arxiv.org/abs/1905.12353v2 (cit. on p. 102).

[179] S. V. Kusminskiy: *Cavity Optomagnonics.* arXiv (2019). URL: http://arxiv.org/abs/1911.11104v1 (cit. on p. 102).

[180] T. Holstein and H. Primakoff: *Field Dependence of the Intrinsic Domain Magnetization of a Ferromagnet.* Physical Review **58**.12 (1940), pp. 1098–1113. DOI: 10.1103/physrev.58.1098 (cit. on p. 103).

[181] J. M. D. Coey: *Magnetism and Magnetic Materials*. Cambridge University Press, 2010. DOI: 10.1017/CBO9780511845000 (cit. on pp. 104, 116).

[182] D. C. Jiles and D. L. Atherton: *Theory of ferromagnetic hysteresis (invited)*. Journal of Applied Physics **55**.6 (1984), pp. 2115–2120. DOI: 10.1063/1.333582 (cit. on pp. 105, 108).

[183] Y. Tabuchi, S. Ishino, T. Ishikawa, R. Yamazaki, K. Usami, and Y. Nakamura: *Hybridizing Ferromagnetic Magnons and Microwave Photons in the Quantum Limit*. Physical Review Letters **113**.8 (2014). DOI: 10.1103/physrevlett.113.083603 (cit. on pp. 106, 121).

[184] M. Harder and C.-M. Hu: "Cavity Spintronics: An Early Review of Recent Progress in the Study of Magnon-Photon Level Repulsion". *Solid State Physics*. Elsevier, 2018, pp. 47–121. DOI: 10.1016/bs.ssp.2018.08.001 (cit. on p. 106).

[185] R. G. E. Morris, A. F. van Loo, S. Kosen, and A. D. Karenowska: *Strong coupling of magnons in a YIG sphere to photons in a planar superconducting resonator in the quantum limit*. Scientific Reports **7**.1 (2017). DOI: 10.1038/s41598-017-11835-4 (cit. on pp. 107, 122).

[186] A. R. Bedyukh, V. V. Danilov, A. Y. Nechiporuk, and V. F. Romanyuk: *Resonant magnetic properties of gadolinium–gallium garnet single crystals*. Low Temperature Physics **25**.3 (1999), pp. 182–183. DOI: 10.1063/1.593724 (cit. on p. 107).

[187] A. A. Serga, A. V. Chumak, and B. Hillebrands: *YIG magnonics*. Journal of Physics D: Applied Physics **43**.26 (2010), p. 264002. DOI: 10.1088/0022-3727/43/26/264002 (cit. on p. 108).

[188] T. Piskor: *Hybrid Quantum Circuits of Microwave Resonators and Thin Film Ferromagnets*. MA thesis. Karlsruhe Institute of Technology, 2018 (cit. on pp. 108, 109, 117).

[189] J. T. Hou and L. Liu: *Strong Coupling between Microwave Photons and Nanomagnet Magnons*. Physical Review Letters **123**.10 (2019). DOI: 10.1103/physrevlett.123.107702 (cit. on pp. 108, 122).

[190] Y. Li, T. Polakovic, Y.-L. Wang, J. Xu, S. Lendinez, Z. Zhang, J. Ding, T. Khaire, H. Saglam, R. Divan, J. Pearson, W.-K. Kwok, Z. Xiao, V. Novosad, A. Hoffmann, and W. Zhang: *Strong Coupling between Magnons and Microwave Photons in On-Chip Ferromagnet-Superconductor Thin-Film Devices*. Physical Review Letters **123**.10 (2019). DOI: 10.1103/physrevlett.123.107701 (cit. on pp. 108, 122).

[191] W. Pepperhoff and M. Acet: *Constitution and Magnetism of Iron and its Alloys*. Springer Berlin Heidelberg, 2001. DOI: 10.1007/978-3-662-04345-5 (cit. on p. 108).

[192] M. A. W. Schoen, D. Thonig, M. L. Schneider, T. J. Silva, H. T. Nembach, O. Eriksson, O. Karis, and J. M. Shaw: *Ultra-low magnetic damping of a metallic ferromagnet*. Nature Physics **12**.9 (2016), pp. 839–842. DOI: 10.1038/nphys3770 (cit. on pp. 108, 117).

[193] M. Tavis and F. W. Cummings: *Exact Solution for anN-Molecule—Radiation-Field Hamiltonian*. Physical Review **170**.2 (1968), pp. 379–384. DOI: 10.1103/physrev.170.379 (cit. on p. 109).

[194] H. Huebl, C. W. Zollitsch, J. Lotze, F. Hocke, M. Greifenstein, A. Marx, R. Gross, and S. T. B. Goennenwein: *High Cooperativity in Coupled Microwave Resonator Ferrimagnetic Insulator Hybrids*. Physical Review Letters **111**.12 (2013). DOI: 10.1103/physrevlett.111.127003 (cit. on pp. 110, 122).

[195] A. Imamoğlu: *Cavity QED Based on Collective Magnetic Dipole Coupling: Spin Ensembles as Hybrid Two-Level Systems*. Physical Review Letters **102**.8 (2009). DOI: 10.1103/physrevlett.102.083602 (cit. on p. 114).

[196] P. Winkel, K. Borisov, L. Grünhaupt, D. Rieger, M. Spiecker, F. Valenti, A. V. Ustinov, W. Wernsdorfer, and I. M. Pop: *Implementation of a transmon qubit using superconducting granular aluminum*. arXiv (2019) (cit. on p. 122).

[197] Y. Schön, J. N. Voss, M. Wildermuth, A. Schneider, S. T. Skacel, M. Weides, J. H. Cole, H. Rotzinger, and A. V. Ustinov: *Rabi oscillations in a superconducting nanowire circuit*. npj Quantum Materials **5**.1 (2020). DOI: 10.1038/s41535-020-0220-x (cit. on p. 122).

List of publications

1. **A. Schneider**, J. Braumüller, L. Guo, P. Stehle, H. Rotzinger, M. Marthaler, A. V. Ustinov, and M. Weides: *Local sensing with the multilevel ac Stark effect.* Physical Review A **97**.6 (2018). DOI: `10.1103/PhysRevA.97.062334`.
2. **A. Schneider**, T. Wolz, M. Pfirrmann, M. Spiecker, H. Rotzinger, A. V. Ustinov, and M. Weides: *Transmon qubit in a magnetic field: Evolution of coherence and transition frequency.* Physical Review Research **1**.2 (2019). DOI: `10.1103/physrevresearch.1.023003`.
3. M. Kristen, **A. Schneider**, A. Stehli, T. Wolz, S. Danilin, H. S. Ku, J. Long, X. Wu, R. Lake, D. P. Pappas, A. V. Ustinov, and M. Weides: *Amplitude and frequency sensing of microwave fields with a superconducting transmon qudit.* npj Quantum Information **6**.1 (2020). DOI: `10.1038/s41534-020-00287-w`.
4. M. Pfirrmann, I. Boventer, **A. Schneider**, T. Wolz, M. Kläui, A. V. Ustinov, and M. Weides: *Magnons at low excitations: Observation of incoherent coupling to a bath of two-level systems.* Physical Review Research **1**.3 (2019), p. 032023. DOI: `10.1103/PhysRevResearch.1.032023`.
5. S. Schlör, J. Lisenfeld, C. Müller, A. Bilmes, **A. Schneider**, D. P. Pappas, A. V. Ustinov, and M. Weides: *Correlating Decoherence in Transmon Qubits: Low Frequency Noise by Single Fluctuators.* Physical Review Letters **123**.19 (2019). DOI: `10.1103/physrevlett.123.190502`.
6. Y. Schön, J. N. Voss, M. Wildermuth, **A. Schneider**, S. T. Skacel, M. Weides, J. H. Cole, H. Rotzinger, and A. V. Ustinov: *Rabi oscillations in a superconducting nanowire circuit.* npj Quantum Materials **5**.1 (2020). DOI: `10.1038/s41535-020-0220-x`.
7. T. Wolz, A. Stehli, **A. Schneider**, I. Boventer, R. Macêdo, A. V. Ustinov, M. Kläui, and M. Weides: *Introducing coherent time control to cavity magnon-polariton modes.* Communications Physics **3**.1 (2020). DOI: `10.1038/s42005-019-0266-x`.
8. J. Braumüller, M. Marthaler, **A. Schneider**, A. Stehli, H. Rotzinger, M. Weides, and A. V. Ustinov: *Analog quantum simulation of the Rabi model in the ultra-strong coupling regime.* Nature Communications **8**.1 (2017), p. 779. DOI: `10.1038/s41467-017-00894-w`.

9. J. Braumüller, M. Sandberg, M. R. Vissers, **A. Schneider**, S. Schlör, L. Grünhaupt, H. Rotzinger, M. Marthaler, A. Lukashenko, A. Dieter, A. V. Ustinov, M. Weides, and D. P. Pappas: *Concentric transmon qubit featuring fast tunability and an anisotropic magnetic dipole moment*. Applied Physics Letters **108**.3 (2016), p. 032601. DOI: 10.1063/1.4940230.

Appendix

A Perturbation theory

The full expansion of Eq. (5.6) up to fourth order in perturbation theory is given by the following formulas:

$$
\begin{aligned}
\tilde{E}_k &\approx \tilde{E}_k^0 + \tilde{E}_k^1 + \tilde{E}_k^2 + \tilde{E}_k^3 + \tilde{E}_k^4 \\
\tilde{E}_k^1 &= \langle k| \tilde{H}_\mathrm{R}^1 |k\rangle = 0 \\
\tilde{E}_k^2 &= \sum_{m\neq k} \frac{\left|\langle m| \tilde{H}_\mathrm{R}^1 |k\rangle\right|^2}{\tilde{E}_k^0 - \tilde{E}_m^0} \\
\tilde{E}_k^3 &= \sum_{m\neq k}\sum_{l\neq k} \frac{\langle k| \tilde{H}_\mathrm{R}^1 |m\rangle \langle m| \tilde{H}_\mathrm{R}^1 |l\rangle \langle l| \tilde{H}_\mathrm{R}^1 |k\rangle}{\left(\tilde{E}_k^0 - \tilde{E}_l^0\right)\left(\tilde{E}_k^0 - \tilde{E}_m^0\right)} - \sum_{m\neq k} \underbrace{\langle k| \tilde{H}_\mathrm{R}^1 |k\rangle}_{=0} \frac{\left|\langle k| \tilde{H}_\mathrm{R}^1 |m\rangle\right|^2}{\left(\tilde{E}_k^0 - \tilde{E}_m^0\right)^2} \\
\tilde{E}_k^4 &= \sum_{m\neq k}\sum_{l\neq k}\sum_{p\neq k} \frac{\langle k| \tilde{H}_\mathrm{R}^1 |p\rangle \langle p| \tilde{H}_\mathrm{R}^1 |m\rangle \langle m| \tilde{H}_\mathrm{R}^1 |l\rangle \langle l| \tilde{H}_\mathrm{R}^1 |k\rangle}{\left(\tilde{E}_k^0 - \tilde{E}_l^0\right)\left(\tilde{E}_k^0 - \tilde{E}_m^0\right)\left(\tilde{E}_k^0 - \tilde{E}_p^0\right)} \\
&\quad - \sum_{m\neq k}\sum_{l\neq k} \frac{\left|\langle k| \tilde{H}_\mathrm{R}^1 |m\rangle\right|^2 \left|\langle k| \tilde{H}_\mathrm{R}^1 |l\rangle\right|^2}{\left(\tilde{E}_k^0 - \tilde{E}_m^0\right)^2\left(\tilde{E}_k^0 - \tilde{E}_l^0\right)} + \underbrace{\langle k| \tilde{H}_\mathrm{R}^1 |k\rangle}_{=0} (\dots)
\end{aligned} \tag{1}
$$

B Fabrication parameters of the magnetic sample

Table 1: Magnetic material

adhesion agent		
hotplate	110 °C	50 s
HDMS	in desiccator	2+5 min
resist application		
resist	AZ5214E	
ramp	500 rpm	5 s
spin	6000 rpm	60 s
hotplate	110 °C	50 s
exposure		
dose	$13\,\mathrm{mW\,cm^{-2}}$	2 s
hotplate	110 °C	50 s
flood exposure	$13\,\mathrm{mW\,cm^{-2}}$	30 s
developing		
AZ developer, H_2O, 1 : 1		33 s
sputtering		37 min
Argon pressure	8.5×10^{-3} mbar	
Fe gun power	100 W	
Co gun power	37 W	
stripping		
NEP in ultrasonic bath level 1		5 min

Table 2: Resist pads

adhesion agent		
hotplate	110 °C	50 s
HDMS	in desiccator	2+5 min
resist application		
resist	AZ5214E	
ramp	500 rpm	5 s
spin	6000 rpm	60 s
hotplate	110 °C	50 s
exposure		
dose	$13\,\mathrm{mW\,cm^{-2}}$	2 s
hotplate	110 °C	50 s
flood exposure	$13\,\mathrm{mW\,cm^{-2}}$	30 s
developing		
AZ developer, H_2O, 1 : 1		33 s
glassing		
hotplate	230 °C	6 min
protective coating		
resist	AZ5214E	
ramp	500 rpm	5 s
spin	1000 rpm	60 s
hotplate	110 °C	50 s
dicing		
stripping		
IPA		~20 s
NEP		1 s
IPA		
H_2O		

C Additional measurement data on the coupled system

Additional measurement data on a resonator scan are shown in Fig. 1. For the measurements presented in Fig. 7.8 c) and d) in the main part, we performed an anticrossing fit as included in our software package [141], shown in Fig. 2 and Fig. 3. The results are however only reasonable for the smaller splitting.

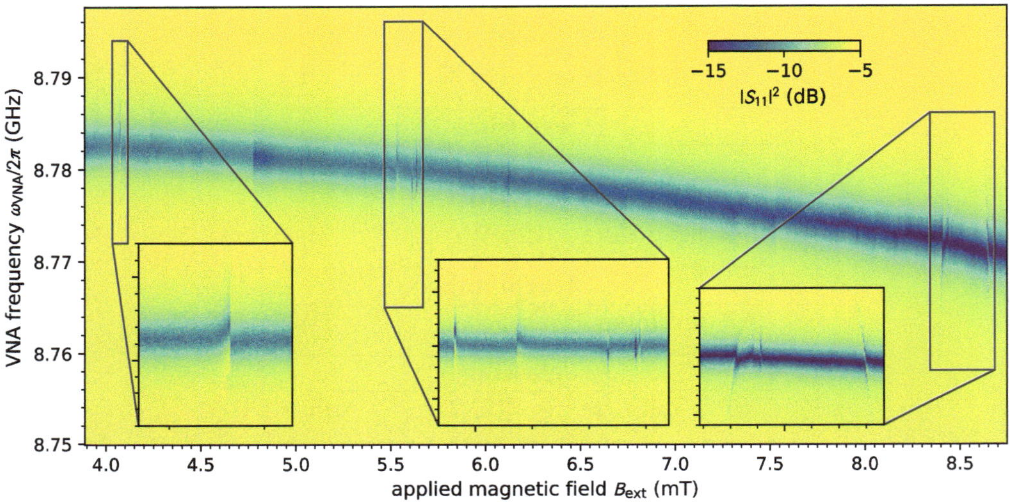

Figure 1: Avoided level crossings in the resonator spectrum. The amplitude of the reflected signal is shown as color against the VNA probe frequency and the applied magnetic field. Data was taken on an upsweep of the magnetic field. The insets show zoom-ins on different features we observed in the resonance spectrum. The experiment was performed similar to Fig. 7.7 and shows a slightly different configuration of resonances.

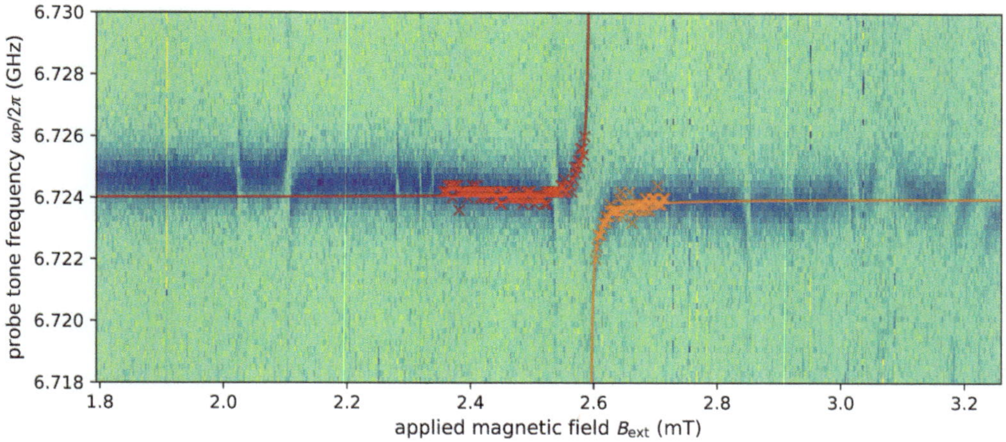

Figure 2: Avoided level crossing in the qubit spectrum. The measurement shown in Fig. 7.8 c) is fitted with an anticrossing fit. Although the fit results seem to accord with the data, the errors on the extracted values are larger than values themselves since the splitting of the curves is large enough that we cannot see both peaks at the same time.

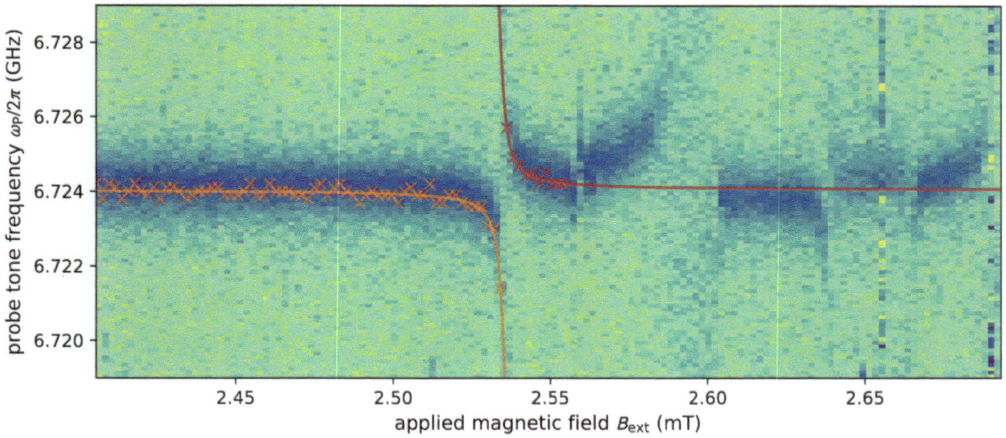

Figure 3: Avoided level crossing in the qubit spectrum. The measurement shown in Fig. 7.8 d) is fitted with an anticrossing fit. From the fit, we find values of $g/2\pi = (3.3 \pm 0.3)\,\mathrm{MHz}$ and a slope of $(3.1 \pm 0.8)\,\mathrm{GHz/mT}$.

Acknowledgements

I want to thank Prof. Martin Weides for giving me the opportunity to work in his "quantum-magnonics" group and perform this thesis under his supervision. We had very fruitful discussions bringing forward the understanding of our data. Despite moving to Glasgow, I never had to wait for a response and encouraging feedback. He also gave me the opportunity to attend various conferences and seminars and encouraged us to order any lab equipment to facilitate our measurements and preparations. Thank you for all the flexibility and freedom I had during my time in your group, which represents the confidence and trust you had in me and my decisions, for which I'm very grateful.

I also would like to thank Prof. Alexey Ustinov for being my second reviewer and for allowing me to be part of his great lab, not only now but also during my Master's thesis. Thank you for your scientific input, the research environment that was created under your management and, of course, for all the group workshops and seminars I was able to attend.

Special thanks goes to Dr. Hannes Rotzinger, who kindly accepted me as officemate, and who was always open for discussions about physics, software, and anything else. Thank you for the incredible knowledge about everything from low temperature physics to software development you shared with me and for all the organization you take care of.

I also want to thank Dr. Christoph Sürgers for helping me with the fabrication of my magnetic samples in his sputtering devices and for his measurements. Christoph was typically the one to ask when nobody else could help, and we borrowed several devices of which nobody (except maybe Hannes) knew that they even exist. Thanks also to Dr. Gerda Fischer for measuring the magnetic samples in her SQUID magnetometer.

All what I did would not have been possible without the help of marvelous colleagues, and I consider myself lucky to have been able to work in such a great research group. My special thanks go to my magnonic colleagues, Dr. Marco Pfirrmann and Tim Wolz, for assembling and maintaining the red fridge, and for our common journey in the "something with magnets" field. Thanks go also to Alex Stehli and Lucas Radtke for their help in the cleanroom with the sample fabrication. I want to thank Dr. Steffen Schlör for the help with and knowledge about any kinds of noise and

Dr. Alexander Bilmes for interesting and mind-broadening discussions about physics. Special thanks go to Dr. Jochen Braumüller for the qubit samples, and especially the great time we had together in the lab and apart from university.

I also would like to thank my colleagues for careful proof-reading of this thesis and other publications. First and foremost I have to mention Tim Wolz here, who never became tired of carefully reading the same text over and over again, while providing valuable and helpful feedback. Also, I'm thanking Dr. Marco Pfirrmann, Max Kristen, Micha Wildermuth and Alex Stehli for carefully reading chapters of this thesis. Huge thanks also to Martin Spiecker for his enthusiasm about the theory behind superconducting qubits and for his explanations and calculations. Your motivation to simply care about physics and not about any political issues is outstanding and unfortunately very rare.

Weiterhin möchte ich mich bei der feinmechanischen Werkstatt bedanken für die präzise Herstellung unserer Probenhalter, Spulenkörper und allem was wir sonst benötigten, sowie für die spontante und unkomplizierte Hilfe bei den „groben Dingen". Besonders bedanke ich mich auch bei der E-Werkstatt, namentlich Roland Jehle, Jannis Ret und Patrick Rust. Danke für die schnelle Hilfe, die gemeinsame Fehlersuche und die Begeisterung für all die kleinen Elektronik-Projekte, die im Labor unglaublich nützlich sind, aber in keiner Arbeit Erwähnung finden. Danke auch an Willi Pensl für die Heliumversorgung und die gemeinsame Arbeit an unserem neuen Verflüssiger.

Mein Dank gilt auch unserem Institutssekretariat, insbesondere Steffi Baatz, die nach Kräften uns Wissenschaftlern den Papierkram vom Hals hält und irgendwie fast immer eine Lösung findet.

Ich bedanke mich bei der Carl Zeiss Stiftung für die finanzielle Förderung während meiner Promotion und die beiden interessanten Stipendiantentreffen.

Während der Arbeit an dieser Promotion war es mir immer eine willkommene Abwechslung, Musik zu machen und gemeinsam etwas auf die Beine zu stellen. Auch wenn es manchmal ein bisschen mehr Stress bedeutete, bin ich froh, dass ich dieses Hobby weiter betreiben konnte. Das unbeschreibliche Gefühl, dass man Menschen begeistern kann und die Bestätigung, die man für ein gemeinsames Engagement erhält, all das sind Erfahrungen, die ich nicht missen möchte. Vielen Dank allen meinen Freunden und Kollegen aus dem Musikverein Büchig, der Guggenmusik und ganz besonders meiner „gloinen Bsetzung", mit denen ich zusammen unglaublich viel Spaß und Freude hatte.

Vielen Dank möchte ich auch meinem Mitbewohner Alex sagen, für die tolle Zeit, die wir zusammen in der WG hatten, für die gemeinsamen Abende und Unternehmungen und nicht zuletzt für die Diskussionen am Frühstückstisch über Mikrowellenelektronik, Quantenphysik und vieles mehr.

den Sand gesteckt hätte. Danke, dass du immer für mich da bist und ich bei dir einfach so sein darf, wie ich bin.

Nicht zuletzt möchte ich mich bei meinen Eltern und meiner Schwester bedanken, ohne die ich diese Arbeit nicht beginnen, durchhalten und vollenden können hätte. Danke, dass ihr mich immer unterstützt und gefördert habt, in den richtigen Momenten an mich geglaubt und an mir gezweifelt habt und dass ihr mich zu dem Menschen gemacht habt, der ich heute bin.

Schade, dass ich nicht mehr mit Oma, Karin und Sabine auf diesen tollen Augenblick anstoßen konnte.

Karlsruhe, Februar 2020 Andre Schneider

Experimental Condensed Matter Physics
(ISSN 2191-9925)

Band 1 Alexey Feofanov
Experiments on flux qubits with pi-shifters. 2011
ISBN 978-3-86644-644-1

Band 2 Stefan Schmaus
Spintronics with individual metal-organic molecules. 2011
ISBN 978-3-86644-649-6

Band 3 Marc Müller
Elektrischer Leitwert von magnetostriktiven Dy-Nanokontakten. 2011
ISBN 978-3-86644-726-4

Band 4 Torben Peichl
Einfluss mechanischer Deformation auf atomare Tunnelsysteme – untersucht mit Josephson Phasen-Qubits. 2012
ISBN 978-3-86644-837-7

Band 5 Dominik Stöffler
Herstellung dünner metallischer Brücken durch Elektromigration und Charakterisierung mit Rastersondentechniken. 2012
ISBN 978-3-86644-843-8

Band 6 Tihomir Tomanic
Untersuchung des elektronischen Oberflächenzustands von Ag-Inseln auf supraleitendem Niob (110). 2012
ISBN 978-3-86644-898-8

Band 7 Lukas Gerhard
Magnetoelectric coupling at metal surfaces. 2013
ISBN 978-3-7315-0063-6

Band 8 Nicht erschienen.

Band 9 Jochen Zimmer
Cooper pair transport in arrays of Josephson junctions. 2014
ISBN 978-3-7315-0130-5

Band 10 Oliver Berg
Elektrischer Transport durch Nanokontakte von Selten-Erd-Metallen. 2014
ISBN 978-3-7315-0209-8

Die Bände sind unter www.ksp.kit.edu als PDF frei verfügbar oder als Druckausgabe bestellbar.

Band 11 Grigorij Jur'evic Grabovskij
Investigation of coherent microscopic defects inside the tunneling barrier of a Josephson junction. 2014
ISBN 978-3-7315-0210-4

Band 12 Cornelius Thiele
STM Characterization of Phenylene-Ethynylene Oligomers on Au(111) and their Integration into Carbon Nanotube Nanogaps. 2014
ISBN 978-3-7315-0235-7

Band 13 Michael Peter Schackert
Scanning Tunneling Spectroscopy on Electron-Boson Interactions in Superconductors. 2014
ISBN 978-3-7315-0238-8

Band 14 Susanne Butz
One-Dimensional Tunable Josephson Metamaterials. 2014
ISBN 978-3-7315-0271-5

Band 15 Philipp Jung
Nonlinear Effects in Superconducting Quantum Interference Meta-Atoms. 2014
ISBN 978-3-7315-0294-4

Band 16 Sebastian Probst
Hybrid quantum system based on rare earth doped crystals. 2016
ISBN 978-3-7315-0345-3

Band 17 Wolfram Kittler
Magnetische Anisotropie und Quantenphasenübergang in $CeTi_{1-x}V_xGe_3$. 2015
ISBN 978-3-7315-0363-7

Band 18 Moritz Peter
Towards magnetic resonance in scanning tunneling microscopy using heterodyne detection. 2015
ISBN 978-3-7315-0410-8

Band 19 Junji Tobias Märkl
Investigation of Magnetic Adatoms with Scanning Tunneling Techniques. 2015
ISBN 978-3-7315-0435-1

Band 20 Jochen Braumüller
Quantum simulation experiments with superconducting circuits. 2018
ISBN 978-3-7315-0780-2

Band 21 Jinjie Chen
Local Investigation of Single Magnetic Molecules with Scanning Tunneling Microscopy. 2018
ISBN 978-3-7315-0819-9

Band 22 Arnold Seiler
Einfluss der Leitungselektronen auf die Dynamik atomarer Tunnelsysteme in ungeordneten Festkörpern: Relaxationsprozesse in metallischen Gläsern und ungeordneten dünnen Aluminiumoxid-Schichten. 2019
ISBN 978-3-7315-0870-0

Band 23 Jasmin Maria Jandke
Elastic and Inelastic Scanning Tunneling Spectroscopy on Iron-Based Superconductors. 2019
ISBN 978-3-7315-0747-5

Band 24 Kevin Edelmann
Electroluminescence from Plasmonic Excitations in a Scanning Tunnelling Microscope. 2019
ISBN 978-3-7315-0923-3

Band 25 Lei Zhang
Sub-Kelvin scanning tunneling microscopy on magnetic molecules. 2019
ISBN 978-3-86644-950-3

Band 26 Lukas Grünhaupt
Granular aluminium superinductors. 2019
ISBN 978-3-7315-0956-1

Band 27 Alexander Bilmes
Resolving locations of defects in superconducting transmon qubits. 2019
ISBN 978-3-7315-0967-7

Band 28 Marco Pfirrmann
Adding nonlinearity to an electromagnetic-magnonic quantum hybrid device. 2020
ISBN 978-3-7315-1003-1

Band 29 Andre Schneider
Quantum Sensing Experiments with Superconducting Qubits. 2021
ISBN 978-3-7315-1032-1